D1085508

MACHINERY ACOUSTICS

MACHINERY ACOUSTICS

GEORGE M. DIEHL

Ingersoll-Rand Company

A WILEY-INTERSCIENCE PUBLICATION

JOHN WILEY & SONS, New York · London · Sydney · Toronto

Copyright © 1973, by John Wiley & Sons, Inc.

All rights reserved. Published simultaneously in Canada.

No part of this book may be reproduced by any means, nor transmitted, nor translated into a machine language without the written permission of the publisher.

Library of Congress Cataloging in Publication Data:

Diehl, George M   1912–
  Machinery acoustics.

  "A Wiley-Interscience publication."
  1. Machinery—Noise. 2. Machinery—Vibration.
3. Machinery—Design. I. Title.

TJ153.D53 1974   620.2′3   73-12980
ISBN 0-471-21360-8

Printed in the United States of America

10 9 8 7 6 5 4 3 2 1

PREFACE

This book deals with machinery noise and how to control it. In many instances it is no more expensive to design a machine to operate quietly than it is to design it to be noisy. In other cases the additional cost of built-in sound control is very slight. The objectives of the book are to acquaint design engineers and those who are responsible for noise reduction with the principles of machinery acoustics, the sources of noise in machinery, and the means available to reduce it. If enough people keep sound control in mind, understand the principles of acoustics, and keep applying them whenever they can, the end products are bound to be quieter.

People responsible for sound control have access to two classes of information: (1) very general articles which define noise as unwanted sound, and say that something should be done about the noise problem, and (2) highly theoretical articles on sound utilizing equations with complex terms the average sound control engineer cannot evaluate. Neither of these articles helps him to solve a practical problem.

Practicing engineers should understand the fundamentals of acoustics, how noise is generated in machinery, which sources of noise are most important, and how machinery noise can be controlled by design.

High-horsepower, high-energy machinery, designed for heavy-duty work, frequently turns out to be noisy, even after great care has been used in design, and it requires additional treatment after it has been built. The design engineer should know how to reduce noise in those instances.

This book is intended to be a guide to the control of machinery noise, and it can be used by technicians as well as by graduate engineers. It contains usable information.

Emphasis is placed on making the subject easy to understand by developing equations to show where they come from and what they mean. Many examples are worked out to show how the equations are used, and each one has a step-by-step procedure.

GEORGE M. DIEHL

Phillipsburg, New Jersey
May 1973

v

CONTENTS

PART TWO   MACHINERY NOISE SOURCES
AND SOUND CONTROL

MACHINERY ACOUSTICS

PART ONE

INTRODUCTION TO SOUND

CHAPTER 1

WAVE MOTION AND SOUND

Sound is produced by a wave motion which is very similar to that of water waves. A motor boat moving in a smooth lake produces a series of waves advancing at a velocity that is quite independent of that of the boat. The waves carry energy since the water in the wave is in motion and therefore has kinetic energy.

The water over which the waves have passed remains calm, and if a floating cork is observed, it is seen to rise and move forward with the crest of the wave, then to sink and move backward in the following trough. This motion is repeated with the next wave, and after the disturbance has passed, the cork will be in its original position. The cork does not follow the wave all the way to the shore, but instead, as each wave passes, it moves forward, then backward, and ends up in its original position. This shows that the water itself is not moved along continuously, but that the motion and the kinetic energy is transferred from one mass of water to the next. Each wave carries with it a definite amount of energy that remains with it as it advances.

1.1 Compressional Waves

The water waves just considered are surface waves, but waves can travel in every direction through an elastic medium, that is, through any medium that has the properties of mass and elasticity.

Mass, or inertia, enables displaced particles to transfer momentum to adjacent particles. Elasticity tends to restore a displaced particle to its original position, like a spring.

Consider a series of masses, like billiard balls, connected together by springs and resting in a frictionless groove (Fig. 1.1):

Figure 1.1

If the first ball is moved toward the second, the spring between them compresses. This compression applies a force to the second ball, which then moves toward the third ball. The next spring is compressed, causing the third ball to move, and so on. The balls are set in motion, one after the other as the wave of compression reaches each spring.

Note that the second ball does not start to move until *after* the first ball has approached it, since no force is exerted on the second ball until the spring is compressed by the motion of the first ball.

This means that the motion of the second ball lags that of the first, and the motion of the third ball occurs still later and lags that of the second.

Similarly, if the first mass is moved away from the second, the first spring is stretched, instead of compressed, causing motion of the second mass, which in turn stretches the second spring. The motion is communicated through the entire series as a wave, accompanied by extension of all the springs.

If the first ball is oscillated back and forth, a series of waves is transmitted with alternating compression and expansion. Each mass oscillates like every other one, but the motion of the second mass always lags that of the first, the third mass lags that of the second, and so on.

This is precisely the kind of wave motion that is set up in air by a tuning fork, or other rapidly vibrating bodies, producing the sensation of sound.

Air has both mass and elasticity. At room temperature and pressure it weighs about 0.075 lb/ft^3. Its elasticity can be demonstrated by pushing a piston into a closed cylinder, for example, a bicycle pump with the air outlet closed. As the air compresses, it offers resistance—just as a spring would. Because air has both of these necessary characteristics it can transmit wave motion.

Whenever an object moves back and forth, or vibrates like the tuning fork, it disturbs the air next to it and causes the air to move back and forth also. The movement produces a variation in normal atmospheric pressure, and the disturbance propagates from particle to particle, causing alternate layers of compression and rarefaction to move away from the vibrating source. When such waves reach the ear of a listener the sensation of "sound" is produced.

It is important to remember that in a compressional wave the particles vibrate longitudinally, or back and forth, in the direction of travel of the wave. There is a phase displacement along the direction of travel, with alternating regions of compression and rarefaction. In the compressed zones the particles are moving forward in the direction of travel, whereas in the rarefied zones they are moving opposite to the direction of wave travel.

1.2 Standing Waves

When a sound wave strikes a wall, or some other object in its path, it is reflected in the same way that light is reflected by a mirror. If the sound wave strikes the wall at some particular angle of incidence, then it is reflected at that same angle. If the wall is perpendicular to the line of travel of the wave, then the reflected wave is returned in the same line as the incident wave, but now it is moving in the opposite direction.

Sound waves reflected from a large flat surface seem to come from a point as far behind the surface as the noise source is in front of it. Echoes from buildings or cliffs are examples of reflected sound.

Both the original sound wave and the reflected one are traveling waves, but they travel in opposite directions. If the original wave is sinusoidal, the reflected one is also sinusoidal. When the two waves meet, a somewhat surprising but commonly encountered effect is produced: the resulting wave no longer travels.

Assume the original sound wave is produced by a piston moving back and forth in a cylinder in simple harmonic motion. If a sensitive pressure transducer capable of detecting instantaneous pressure levels is located at a fixed distance from the source, it indicates all the pressure peaks and valleys in the wave as it moves past the transducer. The pressure increases to a maximum, decreases to zero, goes to a negative maximum, returns again to zero, and then repeats the cycle.

A transducer located slightly farther away from the source goes through the same instantaneous pressure variations, but sees them a little later in time. That is, any sound pressure that exists at one location occurs at a more distant location, but at a later time.

When the original forward-traveling wave and the reflected backward-traveling wave are combined, the resulting wave stands still in space. At each point along the wave the sound pressure increases and decreases with time, but the maximum value of the variation is different at each location, and at certain places it is zero. When this happens the result is called a standing wave. If a microphone is moved along a line between the sound source and the reflecting surface, a series of maxima and minima in sound level is found.

This effect is most pronounced when the noise source produces sinusoidal waves, or discrete frequency components. Complex sounds, composed of many different components, average the effects so that standing waves are not noticeable. However, if they are present and are not handled properly, they can cause large errors in sound measurement.

1.3   Plane Waves and Spherical Waves

The sound waves thus far described were generated by a piston moving back and forth in a tube. If the tube were infinitely long, and there were no reflecting end, the wave would consist of alternating sections of compression and rarefaction, bounded by the walls of the tube. This plane wave, as it is called, is not the only type of sound wave.

Consider a sphere, such as a balloon, expanding and contracting radially in simple harmonic motion. This generates a spherical wave that propagates in the same manner as the plane wave. The speed of propagation is the same as that of the plane wave, and although it now has a spherical front, it is still one dimensional since all of its parameters can be referred to one measurement, the radial distance of the wave front to the center of the sphere.

Many other complex waves can be generated also. For example, a cylindrical tube can vibrate in its radial direction, generating a cylindrical wave. If the ends of the cylinder vibrate in the axial direction at the same time as the cylinder vibrates radially, a more complex two-dimensional wave is produced. Many other types can be imagined.

Nevertheless, in all sound waves there are longitudinal compression and rarefaction. That is, the air is alternately compressed and rarefied in the direction of wave travel. Sound waves differ from light waves in that light consists of transverse waves, or waves that vibrate in a plane normal to the direction of propagation.

It must be remembered that sound is a wave motion, and a knowledge of the behavior of sound waves is important to the proper application of sound control. This must be considered in muffler design, the design of acoustic enclosures, structure-borne sound, and even in the proper placement of sound absorbing material.

CHAPTER 2

PROPERTIES OF SOUND WAVES

2.1 Sound Characteristics

Noise from common sources, such as machinery, is usually complex and is a combination of many sinusoidal components. Each of the components can be described in terms of its *amplitude* and *frequency*. Amplitude determines the loudness of the sound, and frequency determines its pitch.

For example, when a piston is moved back and forth in a cylinder in simple harmonic motion, it produces a variation in atmospheric pressure and generates sound. The distance from the center of the cylinder to one end is called the *amplitude*. The entire length of the cylinder is called double amplitude or *displacement*.

The piston completes one *cycle* when it moves from the center of the cylinder to the far end, reverses its direction, moves to the opposite end of the cylinder, reverses again, and returns to the center of the cylinder.

The number of times per second the piston is moved back and forth through a complete cycle determines the *frequency* of the sound wave—that is, the number of cycles per second (cps). The time required for one complete cycle is called the *period*, and is the reciprocal of the frequency. For example, if the frequency is 1000 cps, the period is 0.001 sec.

Figure 2.1 shows the relations in a sinusoidal wave.

2.2 Velocity of Sound

The speed of sound in air depends on the temperature and is equal to

$$C = 49.03\sqrt{R} \qquad (2.1)$$

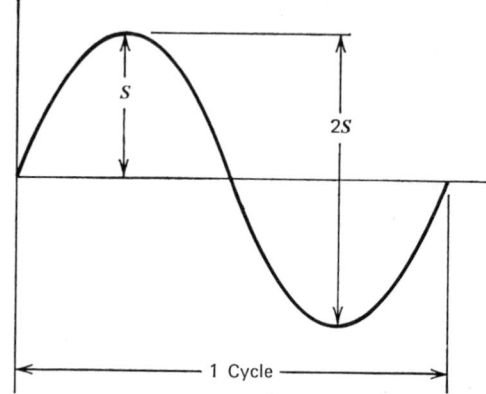

Figure 2.1 S = Amplitude; $2S$ = double amplitude = displacement; f = frequency = number of cycles per second.

where C = speed of sound, in feet per second (fps), and
  R = temperature, in degrees Rankine (459.7 plus the temperature in degrees Fahrenheit).
  For example, at 70°F the speed of sound in air is

$$C = 49.03\sqrt{459.7 + 70}$$

$$= 1128 \text{ fps}$$

The *wavelength* of a sound wave is the distance the wave travels during one cycle. Speed, frequency, and wavelength are related by

$$C = f\lambda \tag{2.2}$$

where C = speed of sound, in feet per second,
  f = frequency, in cycles per second, now called Hertz (Hz), and
  λ = wavelength, in feet.
  At 70°F, when the speed of sound is 1128 fps, the wavelength of a 1000-Hz sound wave is 1.128 ft.
  Sound travels about 4700 fps in water, 16,500 fps in steel, and 13,000 fps in wood. Tables 2.1, 2.2, and 2.3 show the velocity of sound in various solids, liquids, and gases, respectively.
  An understanding of this is important because a different technique must be used for control of long wavelength, low-frequency sound waves than is used for short wavelength, high-frequency sound waves.

TABLE 2.1   VELOCITY OF SOUND IN SOLIDS

Material	Longitudinal Bar Velocity		Plate (bulk) Velocity	
	cm/sec	fps	cm/sec	fps
Aluminum	5.24×10^5	1.72×10^4	6.4×10^5	2.1×10^4
Antimony	3.40×10^5	1.12×10^4	—	—
Bismuth	1.79×10^5	5.87×10^3	2.18×10^5	7.15×10^3
Brass	3.42×10^5	1.12×10^4	4.25×10^5	1.39×10^4
Cadmium	2.40×10^5	7.87×10^3	2.78×10^5	9.12×10^3
Constantan	4.30×10^5	1.41×10^4	5.24×10^5	1.72×10^4
Copper	3.58×10^5	1.17×10^4	4.60×10^5	1.51×10^4
German Silver	3.58×10^5	1.17×10^4	4.76×10^5	1.56×10^4
Gold	2.03×10^5	6.66×10^3	3.24×10^5	1.06×10^4
Iridium	4.79×10^5	1.57×10^4	—	—
Iron	5.17×10^5	1.70×10^4	5.85×10^5	1.92×10^4
Lead	1.25×10^5	4.10×10^3	2.40×10^5	7.87×10^3
Magnesium	4.90×10^5	1.61×10^4	—	—
Manganese	3.83×10^5	1.26×10^4	4.66×10^5	1.53×10^4
Nickel	4.76×10^5	1.56×10^4	5.60×10^5	1.84×10^4
Platinum	2.80×10^5	9.19×10^3	3.96×10^5	1.30×10^4
Silver	2.64×10^5	8.66×10^3	3.60×10^5	1.18×10^4
Steel	5.05×10^5	1.66×10^4	6.10×10^5	2.00×10^4
Tantalum	3.35×10^5	1.10×10^4	—	—
Tin	2.73×10^5	8.96×10^3	3.32×10^5	1.09×10^4
Tungsten	4.31×10^5	1.41×10^4	5.46×10^5	1.79×10^4
Zinc	3.81×10^5	1.25×10^4	4.17×10^5	1.37×10^4
Cork	5.00×10^4	1.64×10^3	—	—
Crystals				
Quartz X cut	5.44×10^5	1.78×10^4	5.72×10^5	1.88×10^4
Rock salt X cut	4.51×10^5	1.48×10^4	4.78×10^5	1.57×10^4
Glass				
Heavy flint	3.49×10^5	1.15×10^4	3.76×10^5	1.23×10^4
Extra heavy flint	4.55×10^5	1.49×10^4	4.80×10^5	1.57×10^4
Heaviest crown	4.71×10^5	1.55×10^4	5.26×10^5	1.73×10^4
Crown	5.30×10^5	1.74×10^4	5.66×10^5	1.86×10^4
Quartz	5.37×10^5	1.76×10^4	5.57×10^5	1.81×10^4
Granite	3.95×10^5	1.30×10^4	—	—
Ivory	3.01×10^5	9.88×10^3	—	—
Marble	3.81×10^5	1.25×10^4	—	—
Slate	4.51×10^5	1.48×10^4	—	—
Wood				
Elm	1.01×10^5	3.31×10^3	—	—
Oak	4.10×10^5	1.35×10^4	—	—

TABLE 2.2   VELOCITY OF SOUND IN LIQUIDS

Material	Temperature °C	°F	Velocity cm/sec	fps
Alcohol, Ethyl	12.5	54.5	1.21×10^5	3.97×10^3
	20	68	1.17×10^5	3.84×10^3
Benzene	20	68	1.32×10^5	4.33×10^3
Carbon Bisulfide	20	68	1.16×10^5	3.81×10^3
Chloroform	20	68	1.00×10^5	3.28×10^3
Ether, Ethyl	20	68	1.01×10^5	3.31×10^3
Glycerine	20	68	1.92×10^5	6.30×10^3
Mercury	20	68	1.45×10^5	4.76×10^3
Pentane	20	68	1.02×10^5	3.35×10^3
Petroleum	15	59	1.33×10^5	4.36×10^3
Turpentine	3.5	38.3	1.37×10^5	4.49×10^3
	27	80.6	1.28×10^5	4.20×10^3
Water, Fresh	17	62.6	1.43×10^5	4.69×10^3
Water, Sea	17	62.6	1.51×10^5	4.95×10^3

TABLE 2.3   VELOCITY OF SOUND IN GASES

Material	Temperature °C	°F	Velocity cm/sec	fps
Air	0	32	3.31×10^4	1.09×10^3
	20	68	3.43×10^4	1.13×10^3
Ammonia Gas	0	32	4.15×10^4	1.48×10^3
Carbon Dioxide	0	32	2.59×10^4	8.50×10^2
Carbon Monoxide	0	32	3.33×10^4	1.09×10^3
Chlorine	0	32	2.06×10^4	6.76×10^2
Ethane	10	50	3.08×10^4	1.01×10^3
Ethyene	0	32	3.17×10^4	1.04×10^3
Hydrogen	0	32	1.28×10^5	4.20×10^3
Hydrogen Chloride	0	32	2.96×10^4	9.71×10^2
Hydrogen Sulfide	0	32	2.89×10^4	9.48×10^2
Methane	0	32	4.30×10^4	1.41×10^3
Nitric Oxide	10	50	3.24×10^4	1.06×10^3
Nitrogen	0	32	3.34×10^4	1.10×10^3
	20	68	3.51×10^4	1.15×10^3
Nitrous Oxide	0	32	2.60×10^4	8.53×10^2
Oxygen	0	32	3.16×10^4	1.04×10^3
	20	68	3.28×10^4	1.08×10^3
Sulfur Dioxide	0	32	2.13×10^4	6.99×10^2
Water Vapor	0	32	1.01×10^4	3.31×10^2
	100	212	1.05×10^4	3.45×10^2

2.3 Sound Power, Sound Intensity, and Sound Pressure

Sound power of a source is the total sound energy radiated by the source per unit of time.

Sound intensity in a specified direction is the sound energy transmitted in that direction through a unit area in a unit of time.

When a sound source radiates power uniformly in all directions, all the power must pass through a spherical surface enclosing the source. If we move away from the source, and the radius of the enclosing sphere is increased, the power per unit area must decrease. The total sound power remains the same, but the enclosing area is increasing, and, therefore, the sound intensity must be decreasing. That is,

$$W = IS \tag{2.3}$$

where W = the total sound power,
   I = the intensity, and
   S = the total surface area.
Figure 2.2 shows this relationship.

At present there are no commercially available instruments that can directly read either sound power or sound intensity. The human ear and microphones respond to sound pressure, and most acoustical instruments are

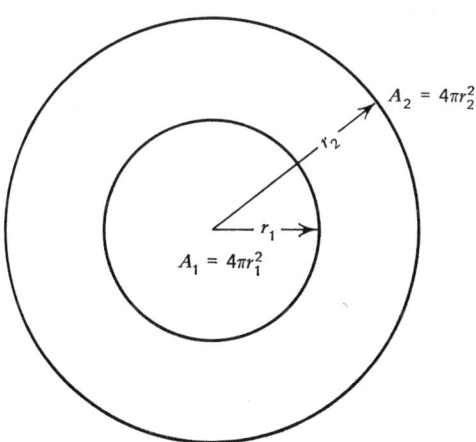

Figure 2.2

calibrated to measure the effective, or root mean square (rms), sound pressure level.

The instantaneous *sound pressure* in a sound wave, measured at any point, is the total instantaneous pressure minus the static pressure at that point. That is, it is the variation from atmospheric pressure.

The effective sound pressure at a point is the rms value of the instantaneous sound pressure over a time interval at the point.

2.4   Decibels and Levels

The term decibel may be thought to be uniquely related to sound. Actually, it originated in electrical engineering. By definition, a decibel is 10 times the logarithm, to the base 10, of a ratio of two powers. For example, if the input power to an electronic amplifier is W_1 and the output power is W_2, the gain of the amplifier in decibels is

$$dB = 10 \log \frac{W_2}{W_1} \qquad (2.4)$$

If W_1 is 1 mW and W_2 is 10 mW, the power gain is

$$dB = 10 \log \tfrac{10}{1}$$

or 10 dB. If the input power is 2 W and the output power is 20 W, the power gain is still 10 dB. Note that the decibel gain does not indicate the actual power developed by the amplifier, nor even what the input or output values actually are. It shows only the ratio of the output to the input. Note also that a 10-dB gain means an increase of 10 to 1, a 20-dB gain means an increase of 100 to 1, and a 30 dB gain, an increase of 1000 to 1. In other words, each time 10 dB are added, the power ratio has been multiplied by 10.

2.4.1   Sound Power Level.   Decibel levels represent only ratios. No information is given concerning actual magnitude, unless a reference level is stated. In acoustics the reference level W_1 has been set at 10^{-12} W. That is,

$$\text{sound power level} = PWL = 10 \log \frac{W}{10^{-12}} \qquad (2.5)$$

If a sound source radiates 10^{-5} W, the sound power *level* is

$$PWL = 10 \log \frac{10^{-5}}{10^{-12}} = 70 \text{ dB} \qquad \text{re } 10^{-12} \text{ W}$$

If it radiates 1.0 W, the sound power *level* is

$$PWL = 10 \log \frac{1.0}{10^{-12}} = 120 \text{ dB} \qquad \text{re } 10^{-12} \text{ W}$$

The range of sound power covered in acoustics is tremendous. A jet engine may radiate 10^{12} times, that is, a million million times the sound power measured in a quiet office. Thus the advantage of the decibel scale is that small numbers can be used in calculations instead of large ones.

Some references, articles on sound, and specifications refer to 10^{-13} W as the reference level for sound power. The new, preferred reference is 10^{-12} W. The reference level should always be clearly stated when discussing sound power level, since it makes a difference of 10 dB. Sound power levels referred to 10^{-13} W are 10 dB higher than those referred to 10^{-12} W. For example, if a machine radiates 0.001 W, its sound power level is

$$10 \log \frac{0.001}{10^{-12}} = 90 \text{ dB} \qquad \text{re } 10^{-12} \text{ W}$$

and

$$10 \log \frac{0.001}{10^{-13}} = 100 \text{ dB} \qquad \text{re } 10^{-13} \text{ W}$$

2.4.2 Sound Pressure Level. In electrical engineering it can be shown that when the impedance is constant, the power ratio is equal to the square of the voltage ratio. That is,

$$\frac{W_2}{W_1} = \left(\frac{E_2}{E_1}\right)^2$$

where E_2 and $E_1 =$ the output and input voltages, respectively.

Voltage is electrical pressure and corresponds, in acoustics, to sound pressure. Therefore, for constant acoustical impedance,

$$\frac{W_2}{W_1} = \left(\frac{P_2}{P_1}\right)^2$$

Since power level, in decibels, equals 10 times the logarithm of the power ratio, the same number of decibels would be obtained from 10 times the logarithm of the pressure ratio squared. That is,

$$SPL = 10 \log \left(\frac{P_2}{P_1}\right)^2 = 20 \log \frac{P_2}{P_1} \tag{2.6}$$

where $SPL =$ sound pressure level, in decibels,
$P_2 =$ sound pressure, in newtons per square meter (N/m²), and
$P_1 =$ reference sound pressure, in newtons per square meter.

The reference pressure has been established at 2.0×10^{-5} N/m². It is sometimes written 0.0002 dyne/cm² or 0.0002 μbar, both of which are the same as 2.0×10^{-5} N/m².

Therefore in acoustics,

$$SPL = 20 \log \frac{P}{2.0 \times 10^{-5}} \tag{2.7}$$

The reference pressure was selected many years ago because it was found to be very close to the normal threshold of hearing at a frequency of 1000 Hz.

2.5   Pressure Produced by Sound

By definition a bar is a pressure of 0.9869 atm. A μbar, that is, one-millionth of a bar, equals 1 dyne/cm². Therefore, 1 μbar, or 1 dyne/cm², equals 14.5×10^{-6} psi, and 0.0002 μbar, the reference pressure, equals 29.0×10^{-10} psi.

The sound pressure, in psi, corresponding to any given sound pressure level, in decibels, can be calculated as shown below:

$$dB = 20 \log \frac{P_2}{29.0 \times 10^{-10}}$$

$$\log \frac{P_2}{29.0 \times 10^{-10}} = \frac{dB}{20}$$

$$\frac{P_2}{29.0 \times 10^{-10}} = 10^{dB/20} = \text{antilog} \frac{dB}{20}$$

$$P_2 = 29.0 \times 10^{-10} \times 10^{dB/20} \quad \text{or}$$

$$29.0 \times 10^{-10} \text{ antilog} \frac{dB}{20} \tag{2.8}$$

EXAMPLE 2.1   Find the pressure produced by a sound of 120 dB.

$$P_2 = 29.0 \times 10^{-10} \times 10^{120/20}$$

$$= 29.0 \times 10^{-10} \times 10^6$$

$$= 29.0 \times 10^{-4}$$

$$= 0.0029 \text{ psi}$$

It may be shown by this method that a pressure of 1 atm corresponds to 194 dB.

2.6 Adding Decibels

It is often necessary to combine sound levels from several sources. For example, it may be desired to estimate the combined effect of adding another machine in an area where other equipment is operating. The procedure for doing this is to combine the sounds on an energy basis.

EXAMPLE 2.2 Assume that three sounds of different frequencies are to be combined to obtain the total sound pressure level. Let the three sound pressure levels be (a) 90 dB, (b) 88 dB, and (c) 85 dB.

Solution:

$$dB = 10 \log \left(\frac{P_2}{P_1}\right)^2 \quad \text{or} \quad \left(\frac{P_2}{P_1}\right)^2 = \text{antilog} \frac{dB}{10} \tag{2.6}$$

Therefore, 1. Mean square sound pressure ratio of

$$(a) = \text{antilog} \frac{90}{10} = 10.0 \times 10^8$$

2. Mean-square sound pressure ratio of

$$(b) = \text{antilog} \frac{88}{10} = 6.31 \times 10^8$$

3. Mean-square sound pressure ratio of

$$(c) = \text{antilog} \frac{85}{10} = 3.16 \times 10^8$$

4. Total mean-square sound pressure ratio $= (10.0 \times 10^8) +$

$$(6.31 \times 10^8) + (3.16 \times 10^8) = 19.47 \times 10^8$$

$$10 \log (19.47 \times 10^8) = 92.9 \text{ dB}$$

An easier method of doing this is shown in Fig. 2.3. When using the chart any two sounds are combined and the sum added to the next sound. It can be seen that when the 90- and 88-dB sounds are combined, the difference of 2 dB indicates that 2.1 dB should be added to the higher level, making 92.1 dB. Then this is combined with the 85-dB sound. The difference is now 7.1 dB, and the number of decibels to be added to the 92.1 dB is 0.79, making the total 92.9, which agrees with that calculated above.

The curve can be plotted from data obtained from the following calculation:

$$dB_1 = 10 \log \frac{W_1}{W_0} \quad \text{and} \quad dB_2 = 10 \log \frac{W_2}{W_0}$$

where $W_1 =$ the sound power, in watts, of the first sound,
$W_2 =$ the sound power, in watts, of the second sound, and
$W_0 =$ the reference level of 10^{-12} W.

$$\frac{dB_1}{10} = \log \frac{W_1}{W_0} \qquad \frac{dB_2}{10} = \log \frac{W_2}{W_0}$$

$$10^{dB_1/10} = \frac{W_1}{W_0} \qquad 10^{dB_2/10} = \frac{W_2}{W_0}$$

$$W_1 = W_0 \cdot 10^{dB_1/10} \qquad W_2 = W_0 \cdot 10^{dB_2/10}$$

$$W_T = W_1 + W_2$$

$$\frac{W_1 + W_2}{W_0} = 10^{dB_1/10} + 10^{dB_2/10}$$

$$dB_T = 10 \log \frac{W_1 + W_2}{W_0} = 10 \log [10^{dB_1/10} + 10^{dB_2/10}]$$

$dB_T - dB_1 =$ the correction factor to be added to dB_1 to obtain dB_T.

$$dB_T - dB_1 = 10 \log [10^{dB_1/10} + 10^{dB_2/10}] - dB_1$$

Let $X = dB_1 - dB_2$. Then

$$dB_2 = dB_1 - X$$

and

$$\begin{aligned}
dB_T - dB_1 &= 10 \log [10^{dB_1/10} + 10^{dB_1-X/10}] - dB_1 \\
&= 10 \log [10^{dB_1/10} + 10^{dB_1/10}(10^{-X/10})] - dB_1 \\
&= 10 \log [10^{dB_1/10}(1 + 10^{-X/10})] - dB_1 \\
&= 10 \log 10^{dB_1/10} + 10 \log (1 + 10^{-X/10}) - dB_1 \\
&= dB_1 + 10 \log (1 + 10^{-X/10}) - dB_1 \\
&= 10 \log (1 + 10^{-X/10}) \qquad\qquad (2.9)
\end{aligned}$$

X	$\dfrac{-X}{10}$	$10^{-X/10}$	$1 + 10^{-X/10}$	$10 \log (1 + 10^{-X/10})$
0	0	1.000	2.000	3.01
1	−0.1	0.794	1.794	2.54
2	−0.2	0.631	1.631	2.12
3	−0.3	0.501	1.501	1.76
4	−0.4	0.398	1.398	1.46
5	−0.5	0.316	1.316	1.19
6	−0.6	0.251	1.251	0.97
7	−0.7	0.199	1.199	0.79
8	−0.8	0.158	1.158	0.64
9	−0.9	0.126	1.126	0.51
10	−1.0	0.100	1.100	0.41

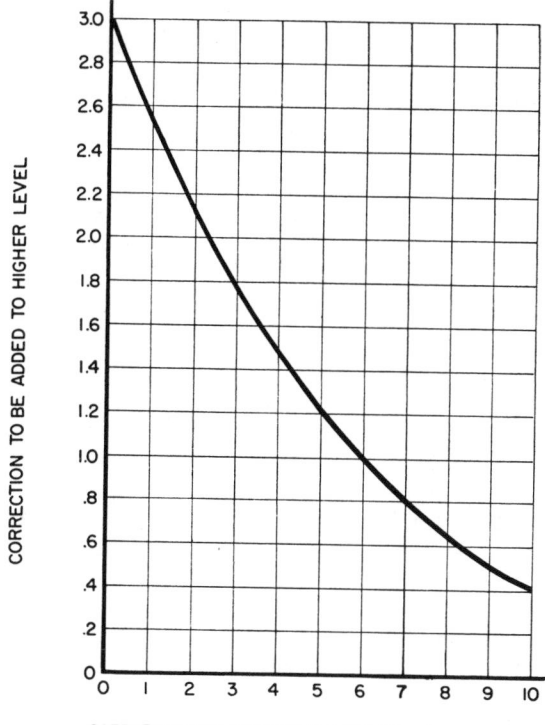

Figure 2.3

Another convenient method for combining decibels is to use a table of antilogs (see Table 2.4). The individual antilogs are listed and added to obtain the antilog of the sum. Ten times the logarithm of this total gives the decibel level of the combined readings.

EXAMPLE 2.3   Find the sum of 95, 90, 87, 85, and 80 dB.

Decibels	Antilog $\dfrac{dB}{10}$
95	31.62×10^8
90	10.00×10^8
87	5.01×10^8
85	3.16×10^8
80	1.00×10^8
Total	50.79×10^8

$10 \log 50.79 \times 10^8 = 97$ dB

TABLE 2.4

dB	Antilog $\dfrac{dB}{10}$	dB	Antilog $\dfrac{dB}{10}$
65	0.03×10^8	91	12.59×10^8
66	0.04×10^8	92	15.85×10^8
67	0.05×10^8	93	19.95×10^8
68	0.06×10^8	94	25.12×10^8
69	0.08×10^8	95	31.62×10^8
70	0.10×10^8	96	39.81×10^8
71	0.13×10^8	97	50.12×10^8
72	0.16×10^8	98	63.10×10^8
73	0.20×10^8	99	79.44×10^8
74	0.25×10^8	100	100.00×10^8
75	0.32×10^8	101	125.90×10^8
76	0.40×10^8	102	158.50×10^8
77	0.50×10^8	103	199.50×10^8
78	0.63×10^8	104	251.20×10^8
79	0.79×10^8	105	316.20×10^8
80	1.00×10^8	106	398.10×10^8
81	1.26×10^8	107	501.20×10^8
82	1.59×10^8	108	631.00×10^8
83	2.00×10^8	109	794.40×10^8
84	2.51×10^8	110	1000.00×10^8
85	3.16×10^8	111	1259.00×10^8
86	3.98×10^8	112	1585.00×10^8
87	5.01×10^8	113	1995.00×10^8
88	6.31×10^8	114	2512.00×10^8
89	7.94×10^8	115	3162.00×10^8
90	10.00×10^8		

If these levels are combined two at a time by using the curve in Fig. 2.3, the sum is found to be 97 dB also.

2.7   Combining Sounds of the Same Frequency

The foregoing procedure cannot be used to combine two sounds of the same frequency since the calculation must now include the phase angle between the two sounds. The equation for doing this is the one used to combine any two sinusoidal waves,

$$P = \sqrt{P_1^2 + P_2^2 + 2P_1P_2 \cos (\theta_1 - \theta_2)} \qquad (2.10)$$

where P = the total rms pressure,

$\quad P_1$ = the rms pressure of the first sound,

$\quad P_2$ = the rms pressure of the second sound, and

$\quad (\theta_1 - \theta_2)$ = the phase angle between the two.

If any two sounds have the same frequency they must first be combined using equation (2.10) before using the chart or the calculation method to combine with the other components. Two sounds of equal magnitude and the same frequency can combine to produce 0, or 6 dB greater than one alone, or anything between 0 and 6 dB, depending on the phase angle.

For example, let $P_2 = P_1$, and the two waves be in phase. Then $(\theta_1 - \theta_2) = 0$ and $\cos(\theta_1 - \theta_2) = 1.0$. Since $P_2 = P_1$

$$P = \sqrt{P_1{}^2 + P_1{}^2 + 2P_1{}^2}$$

$$= \sqrt{4P_1{}^2}$$

$$= 2P$$

If $(\theta_1 - \theta_2) = 180°$, $\cos(\theta_1 - \theta_2) = -1.0$ and

$$P = \sqrt{P_1{}^2 + P_1{}^2 - 2P_1{}^2}$$

$$= 0$$

2.8   Sound Produced by Several Machines of the Same Type

The sound produced by a number of machines of the same type can be determined by adding $10 \log N$ to the sound of one machine alone. That is,

$$SPL_{(N)} = SPL + 10 \log N \qquad (2.11)$$

where $SPL_{(N)}$ = sound pressure level of N machines,

$\quad SPL$ = sound pressure level of one machine, and

$\quad N$ = number of machines of the same type.

EXAMPLE 2.4   If one machine radiates 90 dB, what is the sound level of two machines of the same type?

$$SPL_{(2)} = 90 + 10 \log 2$$

$$= 93 \text{ dB}$$

2.9 Subtracting Decibels

When it is necessary to subtract one noise from another, such as when background noise must be subtracted from total noise to obtain the sound produced by a machine alone, either a mathematical procedure similar to that for adding sounds or a curve similar to Fig. 2.3 may be used. The curve for subtracting sounds is different from that for adding them.

The equations for plotting the curve are determined as follows:

$$dB_T = 10 \log \frac{W_T}{W_0} \quad \text{and} \quad dB_2 = 10 \log \frac{W_2}{W_0}$$

where W_T = the sound power, in watts, of the combination of machine and
background,
W_2 = the background sound power, in watts, and
W_0 = the reference power of 10^{-12} W.

$$\frac{dB_T}{10} = \log \frac{W_T}{W_0} \quad \text{and} \quad \frac{dB_2}{10} = \log \frac{W_2}{W_0}$$

$$10^{dB_T/10} = \frac{W_T}{W_0} \qquad\qquad 10^{dB_2/10} = \frac{W_2}{W_0}$$

$$W_T = W_0 \cdot 10^{dB_T/10} \qquad\qquad W_2 = W_0 \cdot 10^{dB_2/10}$$

$$W_1 = W_T - W_2 = W_0 \cdot 10^{dB_T/10} - W_0 \cdot 10^{dB_2/10}$$

where W_1 = sound power, in watts, of the machine alone.

$$\frac{W_1}{W_0} = 10^{dB_T/10} - 10^{dB_2/10}$$

$$dB_1 = 10 \log \frac{W_1}{W_0} = 10 \log [10^{dB_T/10} - 10^{dB_2/10}]$$

$dB_T - dB_1$ = the correction to be subtracted from dB_T to obtain dB_1

$$dB_T - dB_1 = dB_T - 10 \log [10^{dB_T/10} - 10^{dB_2/10}]$$

Let $X = \mathrm{dB}_T - \mathrm{dB}_2$. Then

$$\mathrm{dB}_2 = \mathrm{dB}_T - X$$

$$
\begin{aligned}
\mathrm{dB}_T - \mathrm{dB}_1 &= \mathrm{dB}_T - 10\log\left[10^{\mathrm{dB}_T/10} - 10^{\mathrm{dB}_T - X/10}\right] \\
&= \mathrm{dB}_T - 10\log\left[10^{\mathrm{dB}_T/10} - 10^{\mathrm{dB}_T/10}(10^{-X/10})\right] \\
&= \mathrm{dB}_T - 10\log\left[10^{\mathrm{dB}_T/10}(1 - 10^{-X/10})\right] \\
&= \mathrm{dB}_T - 10\log 10^{\mathrm{dB}_T/10} - 10\log(1 - 10^{-X/10}) \\
&= \mathrm{dB}_T - \mathrm{dB}_T - 10\log(1 - 10^{-X/10}) \\
&= -10\log(1 - 10^{-X/10}) \qquad\qquad (2.12)
\end{aligned}
$$

X	$\dfrac{-X}{10}$	$10^{-X/10}$	$1 - 10^{-X/10}$	$\log(1 - 10^{-X/10})$	$-10\log(1 - 10^{-X/10})$
3	−0.3	0.501	0.499	9.69810 − 10	3.02
4	−0.4	0.398	0.602	9.77960 − 10	2.20
5	−0.5	0.316	0.684	9.83506 − 10	1.65
6	−0.6	0.251	0.749	9.87448 − 10	1.26
7	−0.7	0.199	0.801	9.90363 − 10	0.96
8	−0.8	0.158	0.842	9.92531 − 10	0.75
9	−0.9	0.126	0.874	9.94151 − 10	0.58
10	−1.0	0.100	0.900	9.95424 − 10	0.46

The data above are plotted in Fig. 2.4. Note that the chart should be used only when the total noise exceeds the background noise by 3 dB or more. If the difference is less than 3 dB, a valid sound test cannot be made.

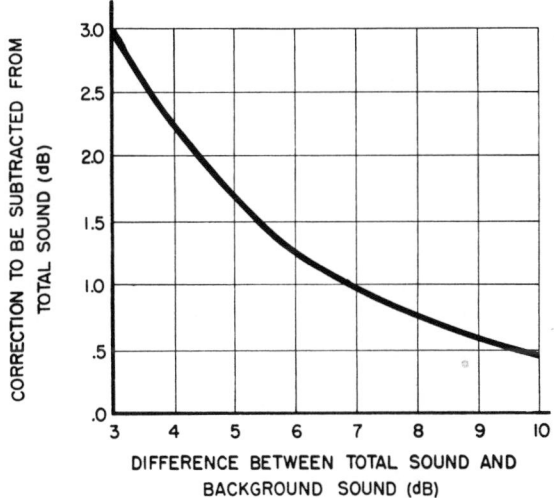

Figure 2.4

2.10   Averaging Decibels

There are many occasions when the average of a number of decibel readings must be calculated. For instance, the sound pressure level at a certain point may be read a number of times, and it is desired to compute the average of the readings.

The procedure for doing this is similar to that for adding decibels, except that the average is found instead of the sum. That is, average $\overline{SPL}$ is calculated by the following equation:

$$\overline{SPL} = 10 \log \frac{1}{n} \left[\text{antilog} \frac{SPL_1}{10} + \text{antilog} \frac{SPL_2}{10} + \cdots \text{antilog} \frac{SPL_n}{10} \right] \quad (2.13)$$

EXAMPLE 2.5   Calculate the average of five sound pressure level measurements if the readings are 90, 87, 87, 88, and 89 dB. From Table 2.4,

Decibels	Antilog $\dfrac{dB}{10}$
90	10.00×10^8
87	5.01×10^8
87	5.01×10^8
88	6.31×10^8
89	7.94×10^8
Total	34.27×10^8

$$\frac{34.27 \times 10^8}{5} = 6.85 \times 10^8$$

$$10 \log 6.85 \times 10^8 = 88.35 \text{ or } 88.4 \text{ dB}$$

Equation (2.13) can be written as

$$\overline{SPL} = 10 \log \left[\text{antilog} \frac{SPL_1}{10} + \text{antilog} \frac{SPL_2}{10} + \cdots \text{antilog} \frac{SPL_n}{10} \right]$$
$$- 10 \log n$$

where $n =$ the number of points.

Therefore, in Example 2.5, $\overline{SPL}$ would be found by taking $10 \log 34.27 \times 10^8$ and subtracting $10 \log 5$. That is,

$$\overline{SPL} = 10 \log 34.27 \times 10^8 - 10 \log 5$$
$$= 10(9.5349) - 10(0.6990)$$
$$= 95.35 - 6.99$$
$$= 88.36 \text{ dB} \quad \text{or} \quad 88.4 \text{ dB}$$

The quantity $10 \log N$ is called the "averaging constant." It is sometimes used in calculations where the average value must be found many times and n is always the same. Mathematically it is exactly the same as equation (2.13).

Although equation (2.13) is the correct way to average decibels, the calculation can be simplified if the maximum variation in levels is small. When the variation is 5 dB or less, the average of the levels can be found by averaging them arithmetically. If the maximum variation is between 5 and 10 dB, the average can be calculated by taking the arithmetic average and adding 1 dB.

For example, in the previous case, where the maximum variation is 3 dB, the arithmetic average gives 88.2 dB which is close enough to the exact value of 88.35.

Let the maximum variation in decibels be between 5 and 10 dB, as in the following example:

Decibels	Antilog $\dfrac{dB}{10}$
90	10.00×10^8
88	6.31×10^8
86	3.98×10^8
85	3.16×10^8
82	1.59×10^8
Total	25.04×10^8

$$\frac{25.04 \times 10^8}{5} = 5.00 \times 10^8$$

$$10 \log 5.00 \times 10^8 = 86.99 \quad \text{or} \quad 87. \text{ dB}$$

Taking the arithmetic average, and adding 1 produces 87 dB also.

2.11   Percent Change in Sound Level

An increase or decrease in sound level can be expressed as a percentage change, instead of a decibel increase or decrease. In general, an expression of percentage change is misleading, even though it is mathematically correct.

A curve, Fig. 2.5, can be plotted to show the relation between dB reduction and percent reduction in sound power. A different curve is needed to show the relation between dB increase and percent increase in sound power.

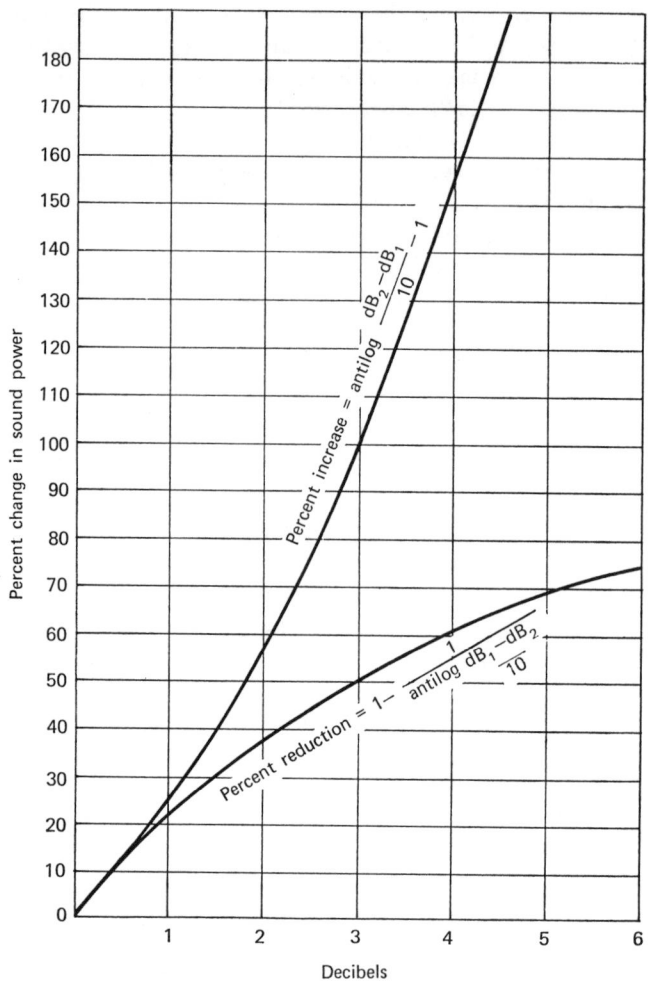

Figure 2.5

Data for the curves can be obtained as follows. Let the sound power level be reduced from dB_1 to dB_2.

$$dB_1 = 10 \log \frac{W_1}{W_0} \quad \text{and} \quad dB_2 = 10 \log \frac{W_2}{W_0}$$

where W_0 = the reference power of 10^{-12} W,
  W_1 = the original power, in watts, and
  W_2 = the final power, in watts.

The reduction in dB level is then

$$dB_1 - dB_2 = 10 \log \frac{W_1}{W_0} - 10 \log \frac{W_2}{W_0}$$

$$= 10 \log \frac{W_1/W_0}{W_2/W_0}$$

$$= 10 \log \frac{W_1}{W_2}$$

Let $X = W_2/W_1$. Then

$$dB_1 - dB_2 = 10 \log \frac{W_1}{XW_1} = 10 \log \frac{1}{X} \qquad (2.14)$$

X	$\dfrac{1}{X}$	$dB = 10 \log \dfrac{1}{X}$	Percent Change
0.10	10.000	10.00	90 (Decrease)
0.20	5.000	6.99	80 (Decrease)
0.30	3.333	5.22	70 (Decrease)
0.40	2.500	3.98	60 (Decrease)
0.50	2.000	3.01	50 (Decrease)
0.60	1.667	2.23	40 (Decrease)
0.70	1.428	1.55	20 (Decrease)
0.80	1.250	0.97	20 (Decrease)
0.90	1.111	0.45	10 (Decrease)
1.00	1.000	0	0 (No Change)
1.10	0.909	0.41	10 (Increase)
1.20	0.833	0.79	20 (Increase)
1.30	0.770	1.14	30 (Increase)
1.40	0.714	1.46	40 (Increase)
1.50	0.667	1.76	50 (Increase)
1.60	0.625	2.04	60 (Increase)
1.70	0.588	2.31	70 (Increase)
1.80	0.555	2.56	80 (Increase)
1.90	0.526	2.79	90 (Increase)
2.00	0.500	3.01	100 (Increase)

Figure 2.5 shows that a 3-dB reduction in sound level means that the original sound power has been reduced by 50 percent, and a 3-dB increase means a 100-percent increase in sound power.

In sound reduction work, a 3-dB improvement is barely noticeable, yet 50 percent of the original sound power has been removed.

Similarly a 10-dB reduction represents a reduction of 90 percent in sound power. Although 10 dB is certainly an appreciable reduction, an observer finds it hard to believe that 90 percent of the original sound power has been removed, and that all he hears is the remaining 10 percent. For this reason, representing noise reduction efforts in terms of percent reduction in sound power should be discouraged.

A similar set of curves can be drawn to express the relation between dB reduction and percent reduction in sound pressure. Here the decrease in decibels can be shown to equal

$$dB_1 - dB_2 = 20 \log \frac{1}{X} \tag{2.15}$$

where $X = $ the ratio of P_2 to P_1.

When the change in decibels is calculated for various values of X, a curve, Fig. 2.6, can be plotted from the following data:

dB	Percent Change
20.00	90 (Decrease)
13.98	80 (Decrease)
10.44	70 (Decrease)
7.96	60 (Decrease)
6.02	50 (Decrease)
4.43	40 (Decrease)
3.10	30 (Decrease)
1.94	20 (Decrease)
0.90	10 (Decrease)
0	0 (No Change)
0.82	10 (Increase)
1.58	20 (Increase)
2.28	30 (Increase)
2.92	40 (Increase)
3.52	50 (Increase)
4.08	60 (Increase)
4.61	70 (Increase)
5.12	80 (Increase)
5.58	90 (Increase)
6.02	100 (Increase)

Note that for any percent change the corresponding decibel change is only one half as large in the case of sound power as it is in the case of sound pressure. Another way of stating this is that for any decibel reduction the

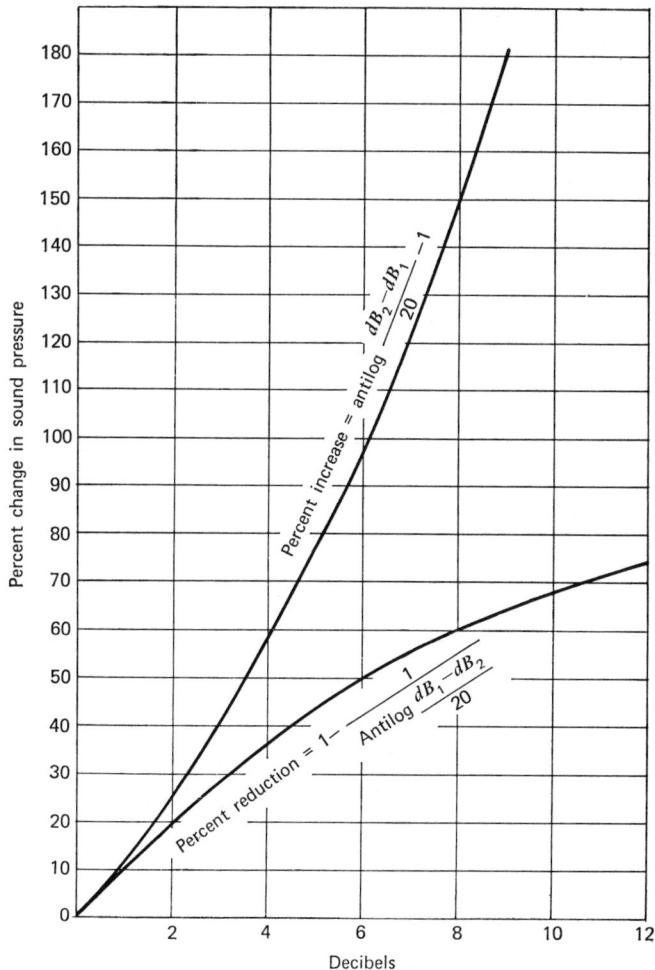

Figure 2.6

percent power change is twice as much as the percent pressure change. This is why some people doing noise reduction work choose to show the results of their efforts in terms of percent reduction in sound power.

A better way to assess a sound reduction is in terms of loudness.

2.12 Phons

A phon is a unit of loudness level. The loudness level of a sound (in phons) is numerically equal to the median sound pressure level (in decibels), relative to $2.0 \times 10^{-5} \, \text{N/m}^2$, of a free progressive wave of frequency 1000 Hz, presented to listeners facing the source, which is judged to be as loud as the sound being rated.

This means that phons are subjective quantities since they are based on the opinions of a group of listeners.

2.13 Sones

As mentioned previously, a good way to decide whether a noise reduction job has accomplished its purpose is on the basis of how much the loudness has been reduced.

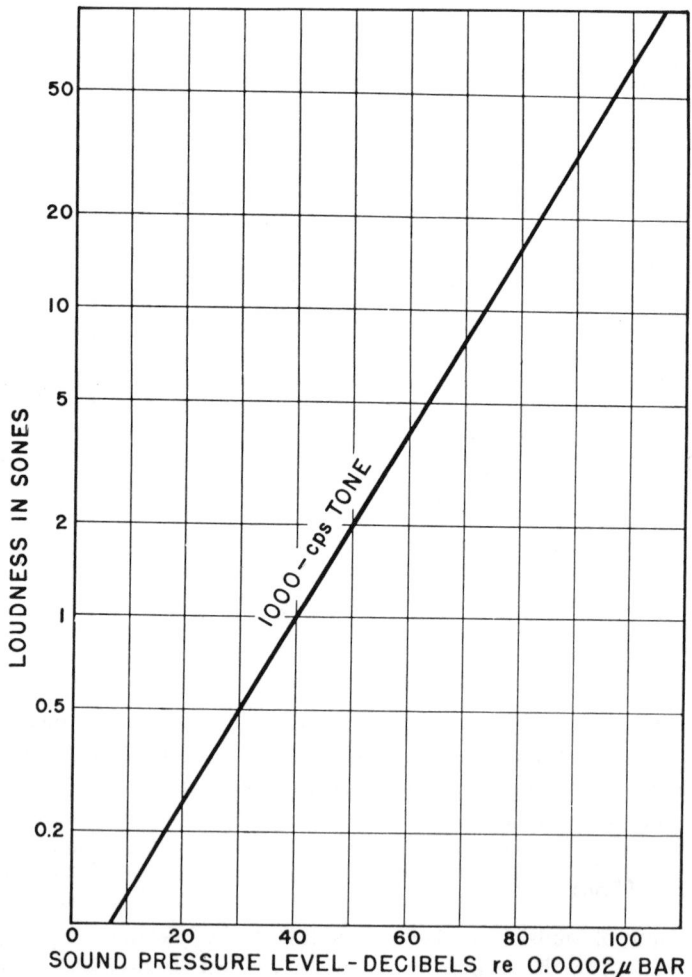

Figure 2.7

Loudness is defined in the *American Standard Acoustical Terminology* as "the intensive attribute of an auditory sensation, in terms of which sounds may be ordered on a scale extending from soft to loud." Listeners are asked to rate sounds according to whether they are half as loud, twice as loud, and so on, as another sound. Loudness depends not only on sound pressure levels, but on frequency as well.

The unit of loudness is the sone. By definition, a simple tone of frequency 1000 Hz, 40 dB above a listener's threshold, produces a loudness of 1 sone. A tone that sounds twice as loud has a loudness of 2 sones. And the loudness of any sound that is judged by the listener to be n times that of the 1-sone tone is n sones.

This relationship is shown on Fig. 2.7. Note that 40 phons = 1 sone, 50 phons = 2 sones, 60 phons = 4 sones, and so on. That is, each time the loudness level is increased by 10 dB, the loudness in sones is doubled. A procedure for calculating loudness level will be described later under the topic "Noise Control Criteria."

μ BAR = dyne/cm²

.0002 μ BAR = average threshold of hearing at 1000Hz

CHAPTER 3

INSTRUMENTATION AND MEASUREMENT

3.1 Sound Level Meter

The basic instrument in all sound measurement is the sound level meter. It consists of a nondirectional microphone, a calibrated attenuator, an amplifier, an indicating meter, and weighting networks. The meter reading is in terms of rms sound pressure level, re 2.0×10^{-5} N/m². Thus the instrument cannot measure the peak level of high-speed sounds, such as those produced by hammer blows, punch presses, or gunshots. Special instrumentation is required for these applications.

The meter usually has two response speeds. FAST response gives an indication correct within 4 dB for a 0.20 to 0.25 sec pulse of 1000 Hz. SLOW response is intended to read the average when meter fluctuations make it difficult to read. The readings on FAST and SLOW are identical, of course, when the sound is steady. Most sound test codes today specify SLOW response.

Usually three weighting networks are included in the instrument—A, B, and C, with frequency response characteristics as shown on Fig. 3.1. These three networks were selected to approximate the response of the human ear at different sound levels. The ear is not as sensitive to low-frequency sounds as it is to high-frequency sounds, and this effect is more pronounced at low-sound levels than at high-sound levels. The original intent of the weighting networks was to use the A-network for sounds below 55 dB, the B-network for sounds between 55 and 85 dB, and the C-network for levels above 85 dB.

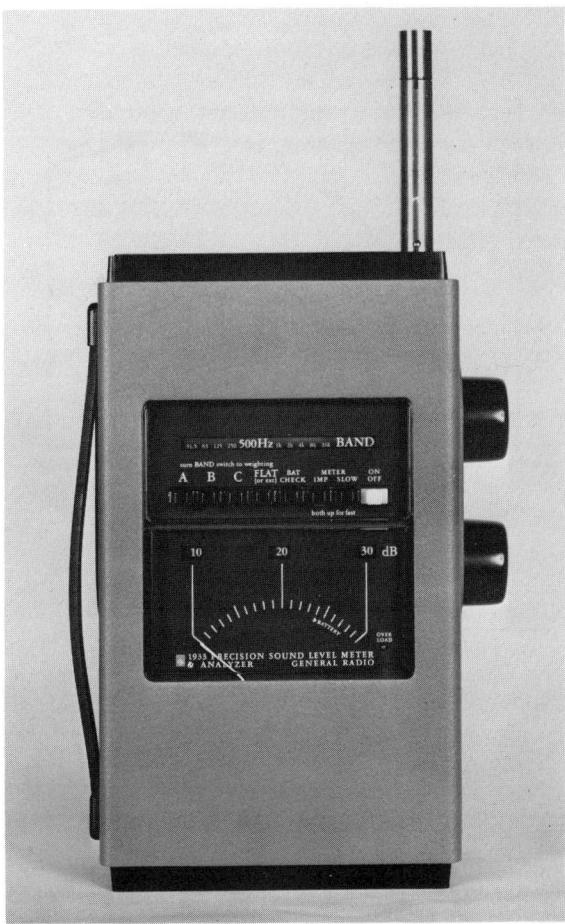

Recent studies have shown that the *A*-network can be used for estimating annoyance caused by noise and for estimating the risk of noise-induced hearing damage. The *B*-network is rarely used today.

An important distinction is made between the numbers read on the *A*- and *C*-scales. The *C*-network is essentially flat, and sounds read with it are called sound pressure levels. The *A*-network falls off sharply at low frequencies to correspond to the response of the ear. Sounds read with it are called sound levels and are referred to as dB*A*. Overall sound can be measured with either the *C*- or *A*-scale, but the most important overall measurement is the one taken on the *A*-scale.

Most new sound level meters have an amplifier with a flat frequency response from 20 to 20,000 Hz, in addition to the usual *A*-, *B*-, and *C*-networks. Overall response when using this network depends on the

Frequency-response characteristics in American Standard
for Sound-Level Meters, S1.4, 1961

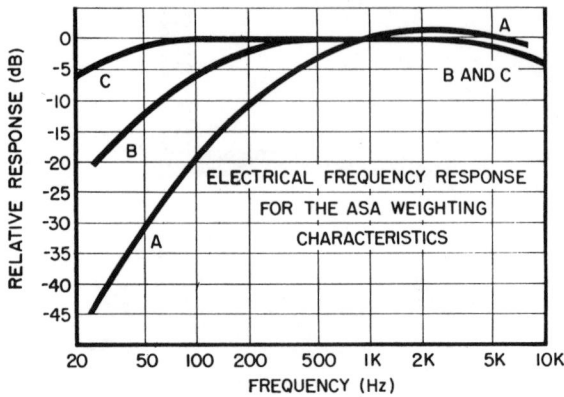

Figure 3.1

microphone being used. All frequency analyses should be made with either the flat or the C-network.

There are two sound level meter standards in use today. *American Standard Specification for General-Purpose Sound Level Meters*, S1.4-1971, and *International Electrotechnical Commission Publication 179, Precision Sound Level Meters* (IEC 179–1965).

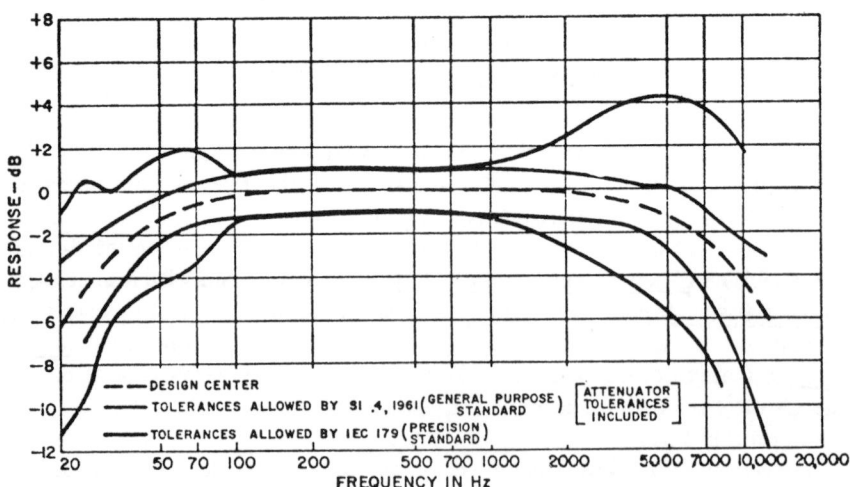

Figure 3.2   C-weighted response tolerances allowed by general-purpose and precision standards.

For *C*-weighting, these two specifications have identical tolerances between 100 and 800 Hz, as shown on Fig. 3.2. Below 100 Hz, the general-purpose specification has closer tolerances than the precision specification, and above 800 Hz, the precision standard is tighter. Also, the general-purpose standard extends only to 10,000 Hz, whereas the precision standard extends to 12,500 Hz.

3.2 Microphones

The microphone is the most important part of the sound measuring instrumentation because the accuracy of the measurement can be no better than that of the microphone. Frequency response, sensitivity, directionality, and range are primarily determined by the microphone. Three general types are available:

1. Crystal microphones of the early type were relatively inexpensive, rugged, and highly sensitive. Their frequency response was not good, however, and they were easily damaged by heat and humidity. Modern crystal microphones retain all the good features of the older models, and, in addition, they have good frequency response. They can operate at temperatures to about 200°F.
2. Dynamic microphones have a coil moving in a magnetic field. They have good frequency response, and can operate at temperatures to 180°F. A disadvantage is that they are affected by magnetic fields and should not be used around motors, generators, or transformers.
3. Condenser microphones have the best frequency response of any today. Some can go as high as 200,000 Hz. Their disadvantages are high cost and the need of an additional preamplifier. Older models were adversely affected by humidity, but the newer models are relatively unaffected.

The noise from most machinery comes from many sources. At any given microphone location the noise comes from different directions, and usually it is considered to be at random incidence with respect to the microphone diaphragm. Therefore, microphones are frequently calibrated for random incident sound. However, other designs can be calibrated for grazing incidence or perpendicular incidence. Care must be taken to use the microphone in the proper orientation, as specified by the manufacturer.

3.2.1 Microphone Sensitivity. One of the most important characteristics of any transducer is its sensitivity. Microphone sensitivity is usually expressed as dBV/μbar, where dBV is the output voltage level in decibels re 1 V for an incident sound pressure wave of 1 μbar (or 0.1 N/m²). Microphone sensitivities are usually in the range of -50 to -125 dBV/μbar. In general, large microphones have high-sensitivity and low-frequency response, whereas small microphones have low-sensitivity and high-frequency response.

If a sound level meter has a microphone with a sensitivity of -50 dBV/μbar, and that microphone is replaced with one having a sensitivity of -60 dBV/μbar, the sound level meter reads 10-dB lower than it did before for the same sound pressure level.

The actual voltage produced by a microphone for any particular sound pressure level can be calculated from its sensitivity in dBV/μbar.

$$dBV = 20 \log \frac{E}{1}$$

$$\frac{dBV}{20} = \log \frac{E}{1}$$

$$E = \text{antilog} \frac{dBV}{20}$$

where E = voltage, in volts, produced by a pressure of 1 μbar.
Also

$$SPL = 20 \log \frac{P}{0.0002}$$

$$\frac{SPL}{20} = \log \frac{P}{0.0002}$$

$$P = 0.0002 \text{ antilog} \frac{SPL}{20}$$

where P = sound pressure corresponding to SPL.
Therefore,

$$E_{out} = \text{antilog} \frac{dBV}{20} \times 2.0 \times 10^{-4} \text{ antilog} \frac{SPL}{20} \tag{3.1}$$

where E_{out} = output voltage, in volts, corresponding to SPL (in decibels, re 0.0002 μbar) when the microphone sensitivity is dBV (in decibels, re 1 V/μbar).

EXAMPLE 3.1 Let the sensitivity = -60 dBV/μbar and the SPL = 90 dB re 0.0002 μbar. Then,

$$E_{out} = \text{antilog} \frac{-60}{20} \times 2.0 \times 10^{-4} \times \text{antilog} \frac{90}{20}$$

$$= \text{antilog} (-3) \times 2.0 \times 10^{-4} \times \text{antilog } 4.5$$

$$= 1 \times 10^{-3} \times 2.0 \times 10^{-4} \times 3.16 \times 10^{+4}$$

$$= 0.006 \text{ V} = 6.0 \text{ mV}$$

3.3   Octave Band Analyzer

The octave-band analyzer is the most common analyzer for industrial noise measurement. As its name implies, it separates the complex noise into frequency bands 1 octave in width, and it measures the level in each of the bands.

An octave is the interval between two sounds having a basic frequency ratio of two. That is, the upper cutoff frequency is twice the lower cutoff frequency. Ideally, the analyzer would pass all frequencies within the octave, and it would have infinite attenuation for all frequencies below the lower cutoff frequency and above the higher cutoff frequency. In practice, this condition does not exist, of course, and the band edge frequencies are the "3-dB down" points, as shown on Fig. 3.3. The filter "skirts," as they are

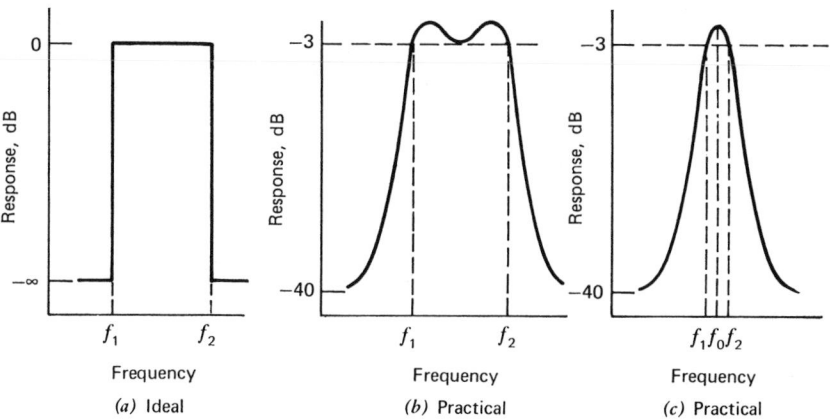

Figure 3.3

called, consist of gradually sloping curves that meet the *American National Standard Specification for Octave, Half-Octave, and Third-Octave Band Filter Sets*, S1.11-1966.

In noise studies, specific frequency bands have been established. The older octaves were 37.5 to 75, 75 to 150, 150 to 300, 300 to 600, 600 to 1200, 1200 to 2400, 2400 to 4800, and 4800 to 9600 Hz. Some instruments had slightly more than an octave at the low and high ends, and these octaves were shown as 20 to 75 and 4800 to 10,000 Hz.

The center frequency of each octave band is its geometric mean, or the square root of the product of the lower and upper cutoff frequencies. That is,

$$f_0 = \sqrt{f_1 f_2} \tag{3.2}$$

where f_0 = center frequency, in Hz,

$\quad f_1$ = lower cutoff frequency, in Hz, and

$\quad f_2$ = upper cutoff frequency, in Hz.

For example, the center frequency of the 75 to 150 Hz octave, is

$$f_0 = \sqrt{75 \times 150} = 106. \text{ Hz}$$

Newer octave band analyzers, instead of identifying the octaves by the lower and upper cutoff frequencies, such as 75 to 150 Hz, identify them by the center frequencies of the new, preferred octave band series. The octaves are now shown as 31.5, 63, 125, 250, 500, 1K (1000), 2K, 4K, and 8K. As before, the upper cutoff frequency is twice the lower cutoff frequency, but the new octaves cover slightly different frequency bands, as shown in Table 3.1.

TABLE 3.1

Old Octaves			New Octaves		
Lower Cutoff	Center Frequency	Upper Cutoff	Lower Cutoff	Center Frequency	Upper Cutoff
37.5	53	75	44	63	88
75	106	150	88	125	177
150	212	300	177	250	354
300	424	600	354	500	707
600	848	1200	707	1K	1414
1200	1697	2400	1414	2K	2828
2400	3394	4800	2828	4K	5656
4800	6787	9600	5656	8K	11312

If equal weight is given to all the frequencies within any octave band, data can be converted from the old octave bands to the new, preferred series, and vice versa, by using the following equations:

New Octaves from Old Octaves   Old Octaves from New Octaves

$$L_{N1} = L_{01}$$
$$L_{01} = 0.42L_{N1} = 0.58L_{N2}$$
$$L_{N2} = 0.68L_{01} + 0.32L_{02}$$
$$L_{02} = 0.19L_{N2} + 0.81L_{N3}$$
$$L_{N3} = 0.68L_{02} + 0.32L_{03}$$
$$L_{03} = 0.19L_{N3} + 0.81L_{N4}$$
$$L_{N4} = 0.69L_{03} + 0.31L_{04}$$
$$L_{04} = 0.18L_{N4} + 0.82L_{N5} \quad (3.3)$$
$$L_{N5} = 0.69L_{04} + 0.31L_{05}$$
$$L_{05} = 0.18L_{N5} + 0.82L_{N6}$$
$$L_{N6} = 0.70L_{05} + 0.30L_{06}$$
$$L_{06} = 0.17L_{N6} + 0.83L_{N7}$$
$$L_{N7} = 0.70L_{06} + 0.30L_{07}$$
$$L_{07} = 0.17L_{N7} + 0.83L_{N8}$$
$$L_{N8} = 0.70L_{07} + 0.30L_{08}$$
$$L_{08} = 0.16L_{N8} + 0.84L_{N9}$$
$$L_{N9} = L_{08}$$

where L_{N1} to L_{N9} = sound pressure levels in the new octaves with center
        frequencies at 31.5, 63, 125 $\cdots$ 8000 Hz, and

   L_{01} to L_{08} = sound pressure levels in the old octave bands with
        lower and upper cutoff frequencies of 20 to 75, 75 to
        150, 150 to 300 $\cdots$ 4800 to 10,000 Hz.

EXAMPLE 3.2   Let the octave band sound pressure level in the 300 to 600 Hz
band be 86 dB, and in the 600 to 1200 Hz band, 105 dB. What is the level in
the new, preferred octave, centered on 500 Hz?

$$L_{N5} = 0.69L_{04} + 0.31\,L_{05}$$

$$= 0.69\,(86) + 0.31\,(105)$$

$$= 59.4 + 32.6$$

$$= 92.\ \ \text{dB}$$

Octave band analyses must be done with either the C-network or the flat
response network. While it is true that the frequency response should be as
flat and uniform as possible to obtain true sound pressure levels, the C-
network is frequently used for this purpose. When instruments are designed
to meet the newest USA and International standards, there is very little
difference between the C-scale and the flat response networks. For example,
in the octave bands centered on 31.5 and 8000 Hz, the C-scale reading is low
by 3 dB. In the 63- and 4000-Hz octaves, the C-scale response is low by
0.8 dB. In all other octaves the difference is negligible.

3.3.1   Overall and dBA Level from Octave Band Data.   The overall sound
pressure level produced by a machine can be calculated by adding the octave
band sound pressure levels by the method explained previously for adding
decibels.

Also the overall dBA level can be calculated from the octave band sound
pressure levels by first applying the A-network correction to the various
bands, and then adding the components.

American National Standard Specification for general-purpose sound level
meters, S1.4-1971, lists the corrections for A-weighting, as shown in Table 3.2.

Octave band analyzers usually provide sufficient information about
industrial machinery to solve noise control problems, but there are some
cases where narrow band analysis is needed.

TABLE 3.2 A-FREQUENCY
WEIGHTING ADJUSTMENTS

f(Hz)	Correction
25	−44.7
31.5	−39.4
40	−34.6
50	−30.2
63	−26.2
80	−22.5
100	−19.1
125	−16.1
160	−13.4
200	−10.9
250	−8.6
315	−6.6
400	−4.8
500	−3.2
630	−1.9
800	−0.8
1,000	0.0
1,250	+0.6
1,600	+1.0
2,000	+1.2
2,500	+1.3
3,150	+1.2
4,000	+1.0
5,000	+0.5
6,300	−0.1
8,000	−1.1
10,000	−2.5

Reference: *American National Standard
Specification for General-Purpose Sound
Level Meters*, S1.4-1971.

3.4 Narrow Band Analyzers

Narrow band analysis is performed when the source of a noise component must be identified for purposes of sound reduction, or some other reason.

Half-octave analyzers have an upper cutoff frequency of $\sqrt{2}$ times the lower cutoff frequency, whereas third-octave analyzers have a ratio of the cube root of 2, or 1.26. That is, the upper cutoff frequency is equal to the cube root of two times the lower cutoff frequency. Tenth-octave analyzers

EXAMPLE 3.3   Calculate the dBA level for the following octave band sound pressure levels:

Octave Band	Band SPL	Band Correction	Corrected SPL	Antilog $\dfrac{dB}{10}$
63	88	−26	62	0.02×10^8
125	88	−16	72	0.16×10^8
250	94	−9	85	3.16×10^8
500	96	−3	93	19.95×10^8
1K	96	0	96	39.81×10^8
2K	92	+1	93	19.95×10^8
4K	89	+1	90	10.00×10^8
8K	76	−1	75	0.32×10^8
				93.37×10^8

$10 \log 93.37 \times 10^8 = 10(9.97) = 99.7 = 100 \text{ dBA}$

have an upper cutoff frequency equal to the tenth root of 2, or 1.07 times the lower cutoff frequency.

Constant-percentage bandwidth analyzers have a bandwidth that is always a fixed percentage of the frequency to which it is tuned. For example, a 2-percent bandwidth analyzer would have a bandwidth of 2 cps when tuned to 100 Hz. Its bandwidth would be 200 cps when tuned to 10,000 Hz.

A constant bandwidth analyzer has a bandwidth that is independent of its setting. A 2-cps bandwidth analyzer always has a bandwidth of 2 cycles/sec, or 2 Hz. This type is difficult to use when there are slight changes in the measured frequency, especially at the high end of the spectrum. For this reason they are not often used for machinery noise analysis.

3.4.1 Analyzer Bandwidth Correction.   When noise is evenly distributed over the spectrum, the level measured by an analyzer with a given bandwidth can be converted to a different bandwidth.

A convenient method for making this calculation is to convert the analyzer reading to spectrum level. This means correcting a reading to an equivalent sound pressure level for a band 1-Hz wide. That is, it is the pressure squared per cycle per second level.

The correction for bandwidth is

$$C = 10 \log \frac{f_2 - f_1}{B} \tag{3.4}$$

where $f_2 - f_1$ = upper frequency minus the lower frequency of the band being measured, and

B = bandwidth where the level is required.

If $B = 1$,

$$C = 10 \log (f_2 - f_1) \tag{3.5}$$

In an octave band the upper cutoff frequency, f_2, is twice the lower cutoff frequency, f_1. That is,

$$f_2 = 2f_1$$

The center frequency is the geometric mean of these two. That is,

$$f_c = \sqrt{f_1 \cdot f_2}$$

Therefore,

$$f_c = \sqrt{f_1 \cdot 2f_1} = \sqrt{2}\, f_1$$

$$f_1 = \frac{f_c}{\sqrt{2}} \quad \text{and since} \quad f_1 = \frac{f_2}{2}$$

$$\frac{f_2}{2} = \frac{f_c}{\sqrt{2}} \quad \text{or} \quad f_2 = \frac{2f_c}{\sqrt{2}}$$

$$\text{Bandwidth} = f_2 - f_1 = \frac{2f_c}{\sqrt{2}} - \frac{f_c}{\sqrt{2}} = \frac{f_c}{\sqrt{2}} = 0.707 f_c$$

In a half-octave band analyzer the upper cutoff frequency is $\sqrt{2}$ times the lower cutoff frequency. That is,

$$f_2 = \sqrt{2}\, f_1$$

The center frequency is the geometric mean of f_1 and f_2, or

$$f_c = \sqrt{f_1 \cdot f_2} = 2^{1/4} \cdot f_1 \quad \text{and} \quad f_1 = \frac{f_c}{2^{1/4}}$$

Also

$$f_2 = 2^{1/4} f_c$$

$$\text{Bandwidth} = f_2 - f_1 = 2^{1/4} f_c - \frac{f_c}{2^{1/4}}$$

$$= f_c \left[2^{1/4} - \frac{1}{2^{1/4}} \right]$$

$$= 0.348 f_c$$

Similarly, in a one-third octave band analyzer the bandwidth is $0.231 f_c$, and in a one-tenth octave band analyzer the bandwidth is $0.0694 f_c$.

The corrections, C, from equation (3.4) can then be listed for the various analyzers as follows:

Bandwidth	Correction
Octave	$10 \log 0.707 f_c$
$\frac{1}{2}$ Octave	$10 \log 0.348 f_c$
$\frac{1}{3}$ Octave	$10 \log 0.231 f_c$
$\frac{1}{10}$ Octave	$10 \log 0.069 f_c$

EXAMPLE 3.4 Calculate the correction to obtain octave band levels from one-third octave band levels.

$$10 \log 0.707 f_c - 10 \log 0.231 f_c = 10 \log \frac{0.707 f_c}{0.231 f_c}$$
$$= 10 \log 3.06$$
$$= 10 \,(0.486)$$
$$= 4.86 \text{ dB}$$

That is, 4.86 dB should be added to the one-third octave band reading to obtain the full octave band reading at the same center frequency.

To compute the one-third octave band reading from the octave band one, 4.86 dB would be subtracted.

3.5 Acoustic Calibrator

An acoustic calibrator is an essential part of a sound measuring system; it fits over the microphone and calibrates the entire system of microphone, attenuator, amplifier, and meter. Various types are available, but a good

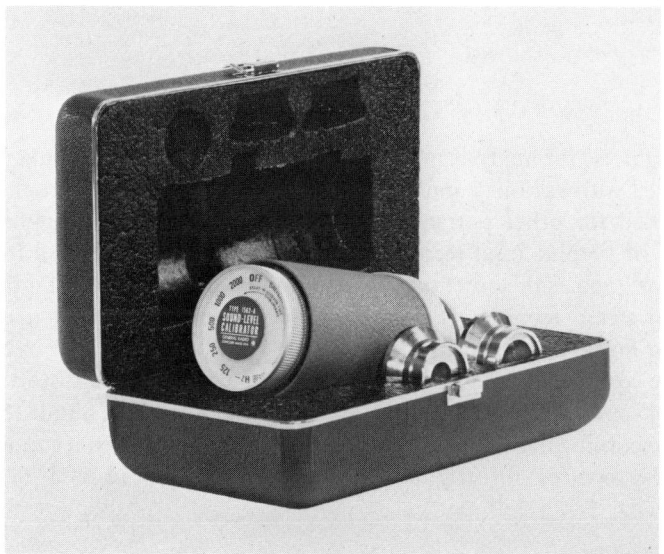

calibrator is one that generates accurate sound pressure levels at several different frequencies. A more elaborate, but expensive, calibrator that can perform absolute calibration of laboratory standard microphones is called a reciprocity calibrator. It is not usually used for industrial sound measurements or field work.

3.6 Wind Screens

Air currents blowing across a microphone produce erratic low-frequency noise because of the turbulence developed on the downstream side. If the wind velocity is high enough, the turbulence noise can exceed the noise being measured and cause errors in the readings. When it is necessary to take data under such conditions, the microphone should be shielded by a wind screen. A suitable one can be made by stretching a thin silk or nylon cloth over a wire frame that completely encloses the microphone. The wind screen should be of fairly large volume and not tight-fitting. About 6 in. in diameter usually works well. Although turbulence noise is still present, the wind screen moves it far enough away from the microphone to prevent it from interfering with the measured noise.

The attenuation of the screen can be measured by reading the same noise, with and without the screen, in a location where there is no wind.

The addition of a wind screen can have an adverse effect on the frequency response of the microphone, and care must be taken to calibrate the system at various frequencies.

Commercially available wind screens made of polyurethane foam are better than those made of silk or nylon, and they can easily be attached to the microphone.

3.7 Recorders

Recording noise for analysis in the laboratory has a definite advantage. It can be analyzed with various bandwidth analyzers, displayed on a graphic chart, and retained for other purposes if desired. However, when magnetic tape recorders or graphic level recorders are used, they should have a frequency response as good as the other components so that distortion is not introduced. The usual stereo recorder is not good enough, even though it may sound perfect for music playback. In addition to flat frequency response over a wide range, the recorder should have low hum and noise level, low distortion, and a constant speed drive. Data should be recorded at a speed of about 15 in./sec.

While recording has definite advantages, it serves best when it supplements, rather than replaces, directly measured data. Directly read data means that

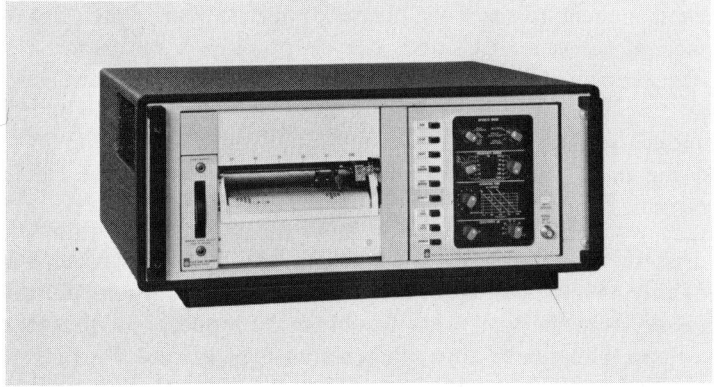

1523 Graphic level recorder with 1/3 octave-band analyzer plug-in (General Radio Company, Concord, Mass.).

you know the answer sooner, and it removes the possible disappointment of having nothing recorded on the tape when you return to the laboratory.

3.8 Microphone Orientation

Microphone placement depends on the objective. If the noise at a worker's ears is to be measured, the microphone should be placed approximately where the worker's ears would be. If a particular noise is to be evaluated, such as that produced by a compressor inlet, then the microphone is located there.

For machinery noise evaluation, the recently developed *ANSI S5.1-1971*, *Test Code for the Measurement of Sound from Pneumatic Equipment* states, "A preliminary survey shall be taken all around the machine at a distance of 1 meter from the nearest major surface of the machine, and at a height of 1.5 meters to locate the point of maximum overall sound level (A scale). This is the primary microphone location. Additional microphone locations shall be established at each end of the plant and at the center of the sides of each casing. All these microphone locations shall be at a horizontal distance of 1 meter from the nearest major surface of the machine, and at a height of 1.5 meters above the floor or walk level."

3.8.1 Data and Sketches. At each of the microphone locations the following data should be taken with the machine operating:

1. Overall sound level using the *A*-weighting network.
2. Octave band sound pressure levels using the flat response or *C*-network. (Overall sound on the *C*-network is not really necessary, but is usually

taken as a check to see if the octave band data look right. Also a comparison of the overall C- and A-levels provides a means for assessing the frequency distribution of the noise.) If the dBA and dBC levels are about the same, the noise contains mostly medium- and high-frequency components. If the dBC level is substantially greater than the dBA level, then the noise can be concentrated in the low-frequency end of the spectrum.

A similar set of data should be taken at one of the locations with the machine shut down to obtain the background noise. Background noise should be subtracted from the total noise to obtain the sound pressure level of the machine alone. This can be done mathematically or graphically, as explained previously, and it should be done on an octave band basis rather than on an overall basis. Remember that if the total noise exceeds the background noise by less than 3 dB, a valid sound test cannot be made.

A sketch should be made showing the machine, locations of other machinery, building walls, and microphone locations. A description of the machine under test should be given and the operating conditions stated.

Test data also should include the make, model, and serial number of the sound test instruments used.

3.9   Attenuator Settings—Dynamic Range

Electrical distortion in a sound measuring system can cause errors in the readings. Overloading of the input amplifier is not uncommon. If the input is too high, the peaks of the incoming sound wave may not be reproduced correctly, so that the wave being analyzed does not have the same shape as the original one, as shown in Fig. 3.4.

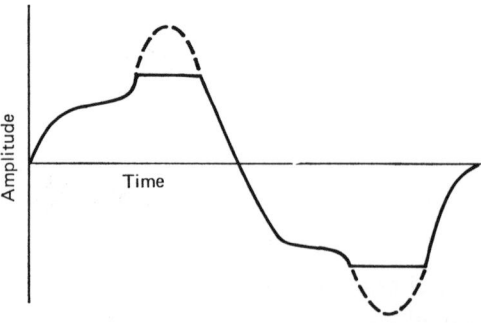

Figure 3.4

If the input amplifier is not overloaded, its own output may be so high that it overloads the following filters, again causing distortion.

If an attempt is made to overcome this by setting the gain of the input amplifier at a low level and compensate for it by increasing the gain of the output amplifier, the electrical circuit noise may be so high that it interferes with the filter output signal (Fig. 3.5).

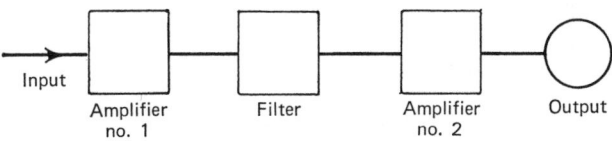

Figure 3.5

The way to test for proper operation is to reduce amplifier gains in steps and see if the output drops proportionately. If it does not, one or more of the amplifiers is being overloaded.

Trouble can also be caused by high-level frequency components that exceed the dynamic range of the filter. For example, if the filter has a dynamic range of 40 dB, a low-frequency component outside the pass band, and 60-dB higher than the frequencies being passed by the filter may still overpower or "swamp out" the filtered band, even after it has been attenuated by 40 dB, as shown in Fig. 3.3.

3.10 Correction for Atmospheric Temperature and Pressure

Atmospheric temperature and pressure affect sound level measurements according to the equation

$$C_1 = 10 \log \left[\left(\frac{460 + t}{528} \right)^{1/2} \left(\frac{30}{B} \right) \right] \tag{3.6}$$

where C_1 = correction, in decibels, to be added to the measured sound pressure level,

t = temperature, in degrees Fahrenheit, and

B = barometric pressure, in inches of mercury.

The relationship is plotted in Fig. 3.6 for various combinations of temperature and pressure.

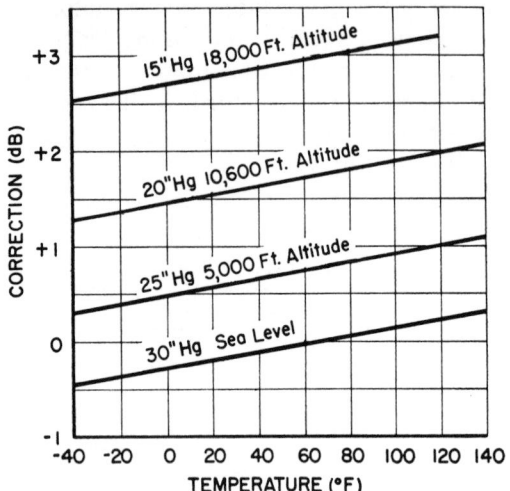

Figure 3.6

EXAMPLE 3.5   Calculate the correction at 20°F and an altitude of 10,600 ft, corresponding to a barometric pressure of 20 in. of mercury.

$$C = 10 \log \left[\left(\frac{460 + 20}{528} \right)^{1/2} \left(\frac{30}{20} \right) \right]$$
$$= 10 \log (0.955 \times 1.5)$$
$$= 10 \log 1.435$$
$$= 10 \times 0.157$$
$$= 1.6 \, dB$$

Therefore, if the measured sound pressure level is 90 dB under these conditions, it would be 91.6 dB when corrected to 68°F and 30 in. of mercury.

In general, sound pressure level measurements are not corrected for the effects of atmospheric temperature and pressure since the corrections are usually not appreciable.

3.11   Correction for Air Attenuation

When sound is propagated through air it is partially absorbed. The absorption is negligible at frequencies below about 1000 Hz, but at higher frequencies it becomes appreciable. It is a function of temperature and humidity, and is linear with distance. One investigation made by Evans and Bazley gives the following: at 68°F and a relative humidity of 60 percent: 0.92 dB/ 1000 ft for 1000 Hz; 2.3 dB/1000 ft for 2000 Hz; 6.4 dB/1000 ft for 4000 Hz; and 24 dB/1000 ft for 8000 Hz.

Figure 3.7 shows the approximate correction for air attenuation for various octave bands and distances.

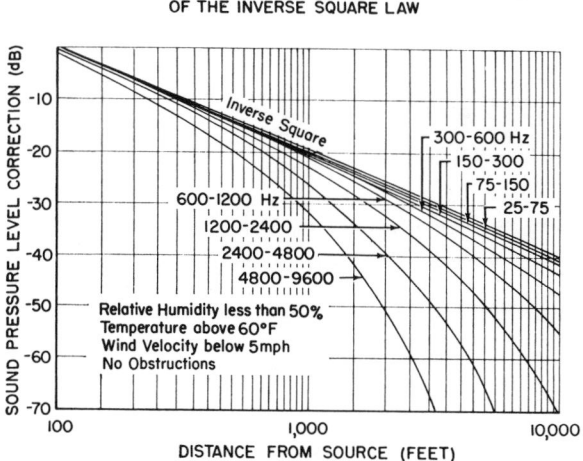

Figure 3.7

3.12   Correction for Wind Effects

Sound waves traveling against the wind tend to be bent upward by the wind. Sound waves traveling in the same direction as the wind tend to be bent downward. This is because whenever there is a wind, there is a wind gradient. Figure 3.8 shows the case where sound waves are traveling against the wind. The wind at the top is blowing faster than it is at the bottom, and, therefore, the sound waves traveling near the bottom are moving faster than those at the top. This bends the sound waves as shown, resulting in less energy at the observer than there would be if no wind were blowing. Conversely, sound waves traveling with the wind are bent downward causing an increase in sound energy near the bottom. This means that the measured sound pressure level downwind is greater than that measured upwind. There is no effect in a direction perpendicular to the direction of the wind.

The magnitude of this effect cannot be calculated simply, and there is very little experimental data on the subject. Theory would seem to indicate that all frequencies should be affected the same way, but tests show otherwise.

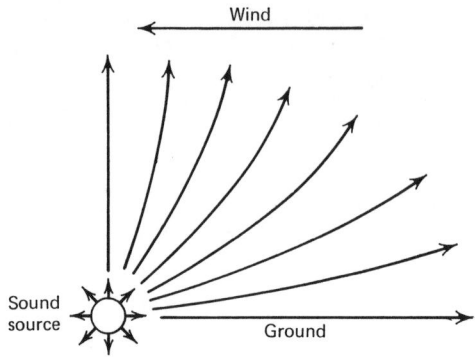

Figure 3.8

Low frequencies seem to be affected very little by the wind, whereas medium- and high-frequency sounds moving against the wind can be attenuated 20 to 30 dB.

3.13   Cable Correction

For greatest accuracy in sound measurement, observers and instruments should be outside the sound field, and only the microphone should be present. In such cases, and in others when it is not desirable to have the observer near the microphone, an extension cable is used to connect the microphone to the sound level meter or analyzer.

When dynamic microphones are used with a matching transformer no correction is needed with cables up to about 100 ft in length.

A correction is usually needed when long cables are used with crystal microphones, and about 7 dB must be added to the sound pressure level reading when the cable length is 25 ft. The exact correction should be furnished by the manufacturer, although it can be measured easily by using a calibrator on the microphone with and without the cable.

3.14   White Noise Generator

In certain measurements it is convenient to use "white noise," or noise that has equal energy per cycle. An octave band analyzer measuring white noise shows a slope of +3 dB/octave. That is, each octave reads 3-dB higher than the preceding octave. This is because each octave admits twice as much sound energy as the previous lower octave. The octave from 300 to 600 Hz has a bandwidth of 300 Hz, whereas the 600 to 1200-Hz octave has a bandwidth of 600 Hz. Since $10 \log 2 = 3$, the slope is 3 dB/octave.

An electrical filter having a slope of -3 dB/octave inserted in the circuit produces a flat response. White noise converted from equal energy per unit frequency to equal energy per octave is called "pink noise."

3.15 Vibration Transducers

Structure-borne noise, or vibration, can be detected and measured by means of displacement, velocity, or acceleration transducers.

Displacement transducers generate an output voltage proportional to the relative displacement between two parts of the pickup. In general, they have poor frequency response and low sensitivity. The output voltage is a function of the displacement only, and it is independent of frequency.

Velocity transducers generate a voltage proportional to the relative velocity of two parts of the pickup. That is, the output is proportional to both displacement and frequency. They are relatively large in size, but have a comparatively high output voltage, and are widely used in various types of vibration measuring systems. Their frequency response is uniform over a wider range than those of displacement pickups.

Accelerometers produce a voltage proportional to the acceleration of the part under study, and, therefore, the output varies directly with the displacement and as the square of the frequency. The upper frequency range of accelerometers is higher than those of displacement or velocity pickups, but at very low frequencies, the output voltage may be small.

3.15.1 Structure-Borne Sound Specifications.
Certain sound specifications are based on structure-borne sound, and there are times when structure-borne sound, or vibration, is more important than airborne sound.

In these cases the specification is based either on velocity or acceleration.

The sound radiated by a machine or structure is usually related to the vibration velocity, and, therefore, maximum permissible levels are stated in terms of "velocity decibels."

As in the case of airborne sound, or anywhere else where decibels are used, a reference level must be stated. Usually the reference level is 10^{-6} cm/sec. That is, the structure-borne sound is stated in terms of "velocity decibels," vdB re 10^{-6} cm/sec.

Because of the better frequency response of accelerometers, most structure-borne vibration specifications are stated in terms of "acceleration decibels," and the reference level is usually 10^{-3} cm/sec². Even when velocity decibels are the primary interest, the specification is stated in terms of acceleration decibels, or adB.

Note that airborne sound, when measured with the A-network, is called dBA, whereas structure-borne sound, in acceleration decibels, is called adB.

3.15.2 adB to vdB. When it is necessary to convert adB re 10^{-3} cm/sec² to the equivalent vdB re 10^{-6} cm/sec, the following calculation can be used:

$$\text{adB} = 20 \log \frac{a}{10^{-3}} \tag{3.7}$$

where adB = acceleration decibels re 10^{-3} cm/sec², and
  a = rms acceleration in cm/sec² rms.

$$\frac{\text{adB}}{20} = \log \frac{a}{10^{-3}}$$

$$\frac{a}{10^{-3}} = 10^{\text{adB}/20}$$

$$a = 10^{\text{adB}/20} \times 10^{-3} = 10^{(\text{adB}/20)-3} \text{ cm/sec}^2 \text{ rms}$$

$$v = \frac{a}{\omega} = \frac{a}{2\pi f}$$

$$v = \frac{10^{(\text{adB}/20)-3}}{2\pi f} \text{ cm/sec rms}$$

$$\text{vdB} = 20 \log \frac{v}{10^{-6}} \tag{3.8}$$

where vdB = velocity decibels re 10^{-6} cm/sec, and
  v = velocity in cm/sec rms.

$$\text{vdB} = 20 \log \frac{10^{(\text{adB}/20)-3}}{2\pi f \times 10^{-6}}$$

$$= 20 \log \frac{10^{(\text{adB}/20)+3}}{2\pi f}$$

$$= 20[\log 10^{(\text{adB}/20)+3} - \log 2\pi f]$$

$$= 20 \left[\frac{\text{adB}}{20} + 3 - 0.798 - \log f \right]$$

$$= \text{adB} + 60. - 15.95 - 20 \log f$$

$$= \text{adB} - 20 \log f + 44$$

or

$$\text{adB} = \text{vdB} + 20 \log f - 44 \tag{3.9}$$

EXAMPLE 3.6 Let adB = 95 re 10^{-3} cm/sec^2 at 900 Hz. Then

$$vdB = 95 - 20 \log 900 + 44$$
$$= 95 - 20 \, (2.95) + 44$$
$$= 95 - 59 + 44$$
$$= 80. \; vdB \; re \; 10^{-6} \; cm/sec$$

3.15.3 adB to FdB. On other occasions, structure-borne sound, or vibration, is stated in terms of force units, and this can be in terms of "force decibels." The reference level is usually 10^{-3} lb, and values are stated in terms of FdB re 10^{-3} lb rms.

If measurements are made in terms of acceleration decibels, and the specification is stated in terms of force decibels, the conversion can be made as follows:

$$adB = 20 \log \frac{a}{10^{-3}} \tag{3.7}$$

where adB = acceleration decibels re 10^{-3} cm/sec^2, and
    a = rms acceleration in cm/sec^2 rms.

$$\frac{adB}{20} = \log \frac{a}{10^{-3}}$$

$$10^{adB/20} = \frac{a}{10^{-3}}$$

$$a = 10^{adB/20} \times 10^{-3} \; cm/sec^2 \; rms$$

$$= 10^{adB/20} \times 10^{-3} \times \sqrt{2} \; cm/sec^2 \; peak$$

$$\frac{a}{g} = 10^{adB/20} \times 10^{-3} \times \sqrt{2} \times \frac{1}{980}$$

where g = acceleration of gravity = 980 cm/sec^2 vector

$$= 10^{adB/20} \times 10^{-3} \times \sqrt{2} \times 1.02 \times 10^{-3}$$

Now,

$$F = W \frac{a}{g}$$

where F = force, in pounds (vector or peak),
    w = weight, in pounds, and
    a/g = acceleration, in gravity units.
Substituting the expression for a/g,

$$F = W \times 10^{adB/20} \times 10^{-3} \times \sqrt{2} \times 1.02 \times 10^{-3} \quad (peak)$$

The force in pounds, rms, is

$$F = W \times 10^{\mathrm{adB}/20} \times 10^{-3} \times \sqrt{2} \times 1.02 \times 10^{-3} \times \frac{1}{\sqrt{2}} \quad \text{(rms)}$$

$$= W \times 10^{\mathrm{adB}/20} \times 1.02 \times 10^{-6} \quad \text{(rms)}$$

By definition,

$$\text{FdB} = 20 \log \frac{F}{10^{-3}} \tag{3.10}$$

where FdB = force decibels re 10^{-3} lb, and
     F = force in pounds, rms.
Therefore,

$$\text{FdB} = 20 \log (W \times 10^{\mathrm{adB}/20} \times 1.02 \times 10^{-3})$$

$$= 20 \log W + \text{adB} + 0.172 - 60$$

$$= \text{adB} + 20 \log W - 59.8 \tag{3.11}$$

EXAMPLE 3.7   If the measured acceleration is 95 adB re 10^{-3} cm/sec^2, and the weight of the part undergoing acceleration is 100 lb. Then

$$\text{FdB} = 95 + 20 \log 100 - 59.8$$

$$= 95 + 40 - 59.8$$

$$= 75.2 \text{ re } 10^{-3} \text{ lb}$$

3.15.4 adB to GdB.   Certain specifications are written in terms of "G-decibels" or GdB. The reference level is usually 10^{-4} G rms.

$$\text{GdB} = 20 \log \frac{G}{10^{-4}} \tag{3.12}$$

where $G = a/g$,
     a = acceleration in cm/sec^2 vector, and
     g = acceleration of gravity, that is, 980 cm/sec^2 vector.
As before,

$$\text{adB} = 20 \log \frac{a}{10^{-3}} \tag{3.7}$$

and

$$a = 10^{\mathrm{adB}/20} \times 10^{-3} \text{ cm/sec}^2 \text{ rms}$$

Therefore,

$$a = 10^{\mathrm{adB}/20} \times 10^{-3} \times \sqrt{2} \text{ cm/sec}^2 \text{ peak}$$

$$\frac{a}{g} = 10^{\mathrm{adB}/20} \times 10^{-3} \times \sqrt{2} \times \frac{1}{980}$$

$$= 10^{\mathrm{adB}/20} \times 10^{-3} \times \sqrt{2} \times 1.02 \times 10^{-3}$$

$$\mathrm{GdB} = 20 \log \frac{a/g}{10^{-4}}$$

$$= 20 \log \left(\frac{a}{g} \times 10^4 \right)$$

$$= 20 \log \left(10^{\mathrm{adB}/20} \times 10^{-3} \times \sqrt{2} \times 1.02 \times 10^{-3} \times 10^4 \right)$$

$$= 20 \log \left(10^{\mathrm{adB}/20} \times \sqrt{2} \times 1.02 \times 10^{-2} \right)$$

$$= 20 \left(\frac{\mathrm{adB}}{20} + 0.15045 + 0.00860 - 2. \right)$$

$$= \mathrm{adB} + 3. + 0.172 - 40$$

$$= \mathrm{adB} - 36.8 \tag{3.13}$$

To Convert Velocity in Inches per Second rms to vdB re 10^{-6} cm/sec rms.
When velocity is measured in terms of inches per second, and the equivalent
vdB is required, the calculation can be made as follows:

$$\mathrm{vdB} = 20 \log \frac{v}{10^{-6}} \tag{3.8}$$

where vdB = velocity decibels re 10^{-6} cm/sec, and
v = velocity in cm/sec rms.
If v is to be measured in inches per second, then

$$\mathrm{vdB} = 20 \log \left(v \times 2.54 \times 10^6 \right)$$
$$= 20 \log v + 8.096 + 120$$
$$= 20 \log v + 128.1 \tag{3.14}$$

EXAMPLE 3.8   Let $v = 0.00125$ in./sec. Then vdB re 10^{-6} cm/sec, is

$$\mathrm{vdB} = 20 \log 0.00125 + 128.1$$
$$= 20(-2.9031) + 128.1$$
$$= -58.06 + 128.1$$
$$= 70.\mathrm{vdB} \text{ re } 10^{-6} \text{ cm/sec}^2$$

CHAPTER 4

SOUND FIELDS

4.1 Free Field

A free sound field is defined by *American Standard Acoustical Terminology* as a homogeneous, isotropic medium free from boundaries. In practice it is a field in which the effects of the boundaries are negligible over the region of interest. It can be outdoors, in a very large room, or in an anechoic chamber. In the latter the walls, floors, and ceiling absorb all the sound incident on them.

A nondirectional source of sound suspended in a free field emits sound uniformly in all directions. As the spherical wave front progresses farther and farther from the source, its area increases with the square of the distance because the area of a sphere is 4π times the square of its radius. Although the sound power remains constant, it is distributed over a larger area, and, therefore, the intensity decreases as the square of the distance from the source. If the distance is doubled, the sound energy spreads over four times the area, and the intensity is one-forth its original value. This is called the "inverse square law."

Sound pressure varies as the square root of the intensity, and, therefore, it decreases directly with distance.

For example, sound pressure level, in decibels, is

$$dB = 20 \log \frac{P}{2.0 \times 10^{-5}} \qquad (2.7)$$

Let the sound pressure be P_1 at distance L_1, and P_2 at distance L_2. Then

$$P_2 = P_1 \times \frac{L_1}{L_2}$$

since sound pressure decreases inversely with distance. Therefore,

$$db_2 = 20 \log \frac{P_2}{2.0 \times 10^{-5}} = 20 \log \frac{P_1 L_1}{2.0 \times 10^{-5} L_2}$$

The difference in decibel level between locations 1 and 2 is

$$dB_1 - dB_2 = 20 \log \frac{P_1}{2.0 \times 10^{-5}} - 20 \log \frac{P_1 L_1}{2.0 \times 10^{-5} L_2}$$

$$= 20 \log \frac{L_2}{L_1}$$

If the distance from microphone to source is doubled, that is, if $L_2 = 2L_1$, then $20 \log 2 = 6$ dB. This means that sound pressure level decreases 6 dB each time the distance to the source is doubled.

The equation can be written as

$$dB_2 = dB_1 - 20 \log \frac{L_2}{L_1} \qquad (4.1)$$

EXAMPLE 4.1   If the sound pressure level at a distance of 20 ft from a machine is 90 dB, calculate the sound pressure level at 70 ft.

$$dB_2 = 90 - 20 \log \frac{70}{20}$$

$$= 90 - 20 \log 3.5$$

$$= 90 - 20 \, (0.544)$$

$$= 79. \text{ dB}$$

4.1.1   Near Field.   Sound pressure level in the vicinity of the usual sound source often shows an appreciable variation with position, even when the source is in a free field. In what is called the "near field," the sound pressure level does not decrease 6 dB each time the distance from the source is doubled. This is because not all parts of the usual sound source vibrate in phase with each other, and there is some cancellation of sound in certain locations. The extent of the near field depends on the frequency of the sound, the dimensions of the source, and the phase relations of the various radiating parts of the source. A frequently used criterion assumes that the near field ends at a distance of about twice the largest dimension of the source. If the source rests on a hard floor, the near field extends to about four times the largest dimension. Beyond that distance, in the "far field," the sound pressure level then decreases 6 dB for each doubling of distance.

4.2   Reverberant Field

Sound from a source in a room with hard walls, floor, and ceiling is reflected back and forth many times. At any particular location in the room the sound is the sum of that which is radiated directly by the source plus all the reflected components coming from many different directions. When a large number of sound waves cross again and again from all directions the sound field is called diffuse. The sound pressure level does not decrease according to the inverse square law, but is uniform throughout the room.

Certain types of sound measurement are performed very conveniently in a reverberant room. For example, sound power level determinations on air-conditioning equipment are usually done in this way.

There are some disadvantages in using reverberant rooms. They are not recommended if the machine under test emits predominantly narrow band sound, or strong discrete frequencies, since they cause standing waves. This, of course, means that the sound energy is not diffuse in the room. The problem can be overcome by using a moving vane to break up the standing wave pattern.

Another disadvantage in using reverberant rooms is that directivity information cannot be obtained.

If the sound source in a reverberant room is stopped suddenly, the sound persists for some time. This persistence is called reverberation.

Airborne sound, incident on a flat surface, behaves in a manner similar to light. Part is reflected and part is absorbed. The reflected component acts like another sound source, and it combines with the original sound, raising the total level. The part that is absorbed has its energy transformed into heating of the absorbing surface. If the walls are very hard and practically all the sound is reflected repeatedly, the sound persists longer than if some of the sound is absorbed each time it strikes a wall.

Reverberation time is defined as the time that would be required for the mean square sound pressure level, originally in a steady state, to decrease 60 dB after the source is stopped. It is measured by suddenly turning off a source of noise in the room and recording the decay rate by means of a microphone, sound level meter, and graphic level recorder. The initial slope of the decay curve can be affected by transients, and the end of the chart can be controlled by the background noise. The slope at the center portion is assumed to be the correct one, and a line is drawn through it and extended. The slope in dB per second is assumed to be constant at this rate, and from this the time to decrease 60 dB is determined. This is the reverberation time.

Figure 4.1 shows the appearance of a typical chart. The vertical axis is in decibels and the horizontal axis is in seconds.

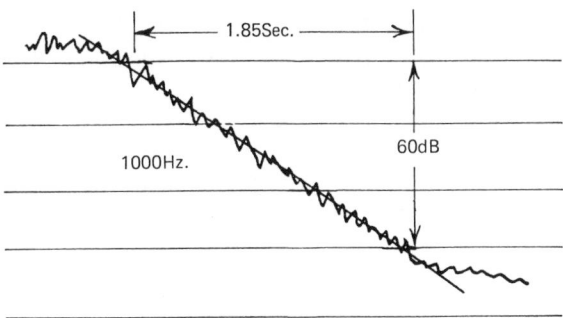

Figure 4.1

Since reverberation time varies with frequency, it must be measured for each frequency, or each frequency band of interest.

4.3   Semireverberant Field

Sound measurements often must be made under actual operating conditions that are neither free field nor reverberant field. That is, the floor, walls, and ceiling are neither completely absorbent nor completely reflecting. Most industrial sound measurements are made in these semireverberant locations.

Sound measurements can be conducted in the same way as they are in a free field, but the room characteristics must be evaluated so that sound pressure level readings can be corrected to their equivalent free-field values.

The characteristics of a semireverberant room are between free field and reverberant field, and are controlled largely by the amount of absorption in the room. If there is a large amount of absorption present, the room is said to have a large room constant, and it behaves in a manner similar to a free field. If there is only a small amount of absorption, the room has a small room constant, and its characteristics are nearer to those of a reverberant room. In practice, all indoor locations are semireverberant, except those specially designed to be highly reverberant and those which have all walls, ceiling, and floor treated with large, highly absorbent wedges. Such rooms are said to be anechoic, and their free-field region extends almost to the boundaries.

If a small source, radiating sound uniformly in all directions, is mounted in the center of a large semireverberant room, the sound pressure level at first decreases with distance from the source almost the same as it does in a free field. It does not follow the inverse square law exactly, but decreases slightly less than 6 dB per doubling of distance.

As the boundaries of the room are approached, the sound pressure level decreases with distance at a lower and lower rate, and it soon remains constant for the remaining distance to the wall.

The directly radiated sound decreases as before, but the sound reflected by the wall increases as the wall is approached so that the total remains about the same. This part of the room is said to be in the reverberant part of the far field, and its extent increases as the amount of absorption decreases.

4.4   Absorption Coefficient

When a sound wave falls on a surface its energy is partially absorbed. The sound-absorbing ability of a material is given in terms of an absorption coefficient designated by α. Absorption coefficient is defined as the ratio of the energy absorbed by the surface to the energy incident on the surface. Therefore, α can be anywhere between 0 and 1. When $\alpha = 0$, all the incident sound energy is reflected; when $\alpha = 1$, all the energy is absorbed.

The value of the absorption coefficient depends on the frequency. Therefore, when specifying the absorbing qualities of a material, either a table or curve showing α as a function of frequency is required. Sometimes a supplier states, for simplicity, the acoustical performance of a material at 500 Hz only, or by a noise reduction coefficient (NRC) that is obtained by averaging, to the nearest multiple of 0.05, the absorption coefficients at 250, 500, 1000, and 2000 Hz.

The absorption coefficient varies somewhat with the angle of incidence of the sound wave. Therefore, for practical use, a statistical average absorption coefficient at each frequency is usually measured and stated by the manufacturer. It is often better to select a sound absorbing material on the basis of its characteristics for a particular noise rather than by its average sound absorbing qualities.

Sound absorption is a function of the length of path relative to the wavelength of the sound, and not the absolute length of the path of sound in the material. This means that at low frequencies the thickness of the material becomes important, and absorption increases with thickness. Low-frequency absorption can be improved further by mounting the material at a distance of one-quarter wavelength from a wall, instead of directly on it.

For example, assume that the frequency to be absorbed is 500 Hz and has a wavelength [from equation (2.2)] of

$$\lambda = \frac{c}{f} = \frac{1128}{500} = 2.25 \text{ ft}$$

One-quarter wavelength is, therefore,

$$\frac{\lambda}{4} = \frac{2.25}{4} = 0.56 \text{ ft}$$

Therefore, the absorption is increased at 500 Hz if the absorbing material is placed approximately 6 in. from the wall, instead of directly on it.

Table 4.1 shows absorption coefficients of various materials used in construction. For comparison, the absorption coefficient of Fiberglas, 6 lb/ft³ density, is also given.

TABLE 4.1 ABSORPTION COEFFICIENTS

Material	125 cps	250 cps	500 cps	1000 cps	2000 cps	4000 cps
Brick, Unglazed	0.03	0.03	0.03	0.04	0.05	0.07
Brick, Unglazed, Painted	0.01	0.01	0.02	0.02	0.02	0.03
Concrete Block	0.36	0.44	0.31	0.29	0.39	0.25
Concrete Block, Painted	0.10	0.05	0.06	0.07	0.09	0.08
Concrete	0.01	0.01	0.015	0.02	0.02	0.02
Wood	0.15	0.11	0.10	0.07	0.06	0.07
Glass, Ordinary Window	0.35	0.25	0.18	0.12	0.07	0.04
Plaster	0.013	0.015	0.02	0.03	0.04	0.05
Plywood	0.28	0.22	0.17	0.09	0.10	0.11
Tile	0.02	0.03	0.03	0.03	0.03	0.02
6 lb/ft³ Fiberglas	0.48	0.82	0.97	0.99	0.90	0.86

The sound absorption of a surface is given in "sabins," which is the absorption of 1 ft² of a perfectly absorbing surface. A surface area, S, that has an absorption coefficient of α, has a total absorption of $S \times \alpha$ sabins.

Average absorption coefficient $\bar{\alpha}$ is calculated as follows:

$$\bar{\alpha} = \frac{\alpha_1 S_1 + \alpha_2 S_2 + \cdots \alpha_n S_n}{S_1 + S_2 + \cdots S_n} \tag{4.2}$$

where

$\bar{\alpha}$ = the average absorption coefficient,

$\alpha_1, \alpha_2, \alpha_n$ = the absorption coefficients of materials on various surfaces, and

S_1, S_2, S_n = the areas of various surfaces.

EXAMPLE 4.2 Calculate the average absorption coefficient, at 1000 Hz, of a room 50-ft long, 30-ft wide, and 15-ft high, if the floor is painted concrete, the ceiling is smooth finish plaster, and the walls are wood paneling.

	Surface Area ft²	Absorption Coefficient	Absorption Sabins
Floor	1500	0.07	105
Ceiling	1500	0.03	45
Side Walls	1500	0.09	135
End Walls	900	0.09	81
	5400		366

$$\bar{\alpha} = \frac{366}{5400} = 0.068$$

4.4.1 Sound Level Reduction due to Increased Absorption in Room. The far-field noise in a room can be reduced by increasing the absorption. The dB reduction can be estimated by the following equation:

$$NR = 10 \log \frac{\bar{\alpha}_2 S}{\bar{\alpha}_1 S} \tag{4.3}$$

where NR = far-field noise reduction, in decibels,
$\bar{\alpha}_1 S$ = room absorption before treatment, and
$\bar{\alpha}_2 S$ = room absorption after treatment.

EXAMPLE 4.3 In Example 4.2 let the side walls be treated with absorbing material that has an absorption coefficient of 0.75. Calculate the noise reduction.

	Surface Area ft²	Absorption Coefficient	Absorption Sabins
Floor	1500	0.07	105
Ceiling	1500	0.03	45
Side Walls	1500	0.75	1125
End Walls	900	0.09	81
	5400		1356

$$NR = 10 \log \frac{1356}{366}$$
$$= 10 \log 3.71$$
$$= 10 \times 0.569$$
$$= 5.7 \text{ dB}$$

Equation (4.3) shows that doubling the absorption reduces the reverberant sound level by 3 dB. This reduction is fairly easy to obtain if the room initially has very little absorption. Subsequent reduction is more difficult to obtain since it must be doubled again to get another 3 dB.

4.5 Room Constant

Room constant is defined by *American Standard Acoustical Terminology* as "the product of the average absorption coefficient of the room and the total internal area of the room divided by the quantity one minus the average absorption coefficient." That is,

$$R = \frac{S_t \bar{\alpha}}{1 - \bar{\alpha}} \qquad (4.4)$$

where R = the room constant, in square feet,
  $\bar{\alpha}$ = the average absorption coefficient, and
  S_t = the total area of the room, in square feet.

In a free field the absorption coefficient is infinity since there is no reflection at all; that is, $R = \infty$. In a room where there is zero absorption, $R = 0$. Of course, neither of these two extreme conditions exists in practice. Increasing the absorption in a room increases the room constant, and decreases the reverberation time. It is obvious that the acoustical characteristics of a room can be described in terms of its room constant.

EXAMPLE 4.4   Calculate the room constant in the 1000-Hz octave band of (a) the room in Example 4.2, and (b) the room in Example 4.3.

$$\text{(a)} \quad R = \frac{S_t \bar{\alpha}}{1 - \bar{\alpha}} \qquad (4.4)$$

$$\bar{\alpha} = \frac{366}{5400} = 0.068 \qquad (4.2)$$

$$R = \frac{5400 \times 0.068}{1 - 0.068} = 395 \text{ ft}^2$$

$$\text{(b)} \quad \bar{\alpha} = \frac{1356}{5400} = 0.251$$

$$R = \frac{5400 \times 0.251}{1 - 0.251} = 1810 \text{ ft}^2$$

4.5.1 Determination of Room Constant from Reverberation Time. Room constant can be determined from reverberation time by the following equation:

$$R = \frac{S}{(TS/0.049V) - 1}$$
(4.5)

where R = the room constant, in square feet,
     T = the reverberation time, in seconds,
     S = the room surface area, in square feet, and
     V = the volume, in cubic feet.

EXAMPLE 4.5   A room whose dimensions are 25-ft long, 15-ft wide, and 10-ft high has a measured reverberant time of 1.5 sec in the 1000-Hz octave. Calculate the room constant.

The total surface area, S, is

$$\text{Floor} = 375$$
$$\text{Roof} = 375$$

Walls $(2 \times 25 \times 10) + (2 \times 15 \times 10) = 800$
$$S = \overline{1550} \text{ ft}^2$$

The room volume is

$$V = 25 \times 15 \times 10 = 3750 \text{ ft}^3$$

Therefore,

$$R = \frac{1550}{(1.5 \times 1550/0.049 \times 3750) - 1}$$
$$= \frac{1550}{12.65 - 1}$$
$$= 133. \text{ ft}^2$$

4.6   Sound Power Level in a Free Field

All vibrating objects radiate sound, and the nature of the sound depends on the type of source and the medium into which the sound is radiated. The source can be described by the following:

1.   The total radiated sound power.
2.   The sound power as a function of frequency, or frequency bands.
3.   The directivity.

The total acoustic power radiated by a source is a fundamental quantity; sound pressure is not. Sound power is independent of the distance between source and microphone, or listener; sound pressure is not. Sound power is not affected appreciably by the kind of room in which the source is placed, but sound pressure is affected considerably.

As an analogy, an ordinary 100-W lamp bulb produces a certain amount of light. The reading on a light meter decreases as an observer walks farther and farther away from the lamp. Similarly, the reading on a sound level meter decreases with distance from the source. Furthermore, the 100-W lamp produces a higher reading on the light meter, at any particular distance, if it is placed in a room with white walls than it does if the walls are painted dull black. Similarly, a sound level meter reads higher when a sound source is placed in a hard walled room than it does if the source is in a free field.

4.6.1 Spherical Radiation.
The total radiated power of a source in a free field can be determined by measuring the sound pressure level at all points on the surface of a hypothetical sphere surrounding the source. The radius of the sphere should be at least twice the largest dimension of the source to avoid near-field effects. In the practical case, measurements can be made only at a fixed number of locations, and the sound pressure levels at other locations must be estimated from these. The error can be minimized by taking a large number of measurements. The sphere over which measurements are made is divided into a number of equal areas, and the power passing through each area is determined in each frequency band of interest. The total power or power level is then calculated for each frequency band by adding the amounts at each microphone location.

4.6.2 20-Point Array.
When greatest accuracy is desired a 20-point microphone array is used. The x-, y-, and z-coordinates of such an array are given in Table 4.2.

The x-, y-, and z-coordinates are measured from the acoustic center of the sound source, which is usually assumed to be at its geometric center. This may not be strictly true, since it may actually be at some other location and may vary with frequency, but in most practical problems the effect of this on the calculated sound power level is negligible.

The microphone should be placed at a radius not less than twice the largest dimension of the machine under test and not less than 2 ft from the nearest major surface of the machine. Furthermore, the microphone should be at least $\lambda/4$ from any wall of the enclosure, where λ is the wavelength of the lowest frequency of interest.

For example, if the lowest frequency of interest is 100 Hz, then the microphone should not be closer to a wall than one-quarter wavelength of a 100-Hz sound wave. This can be calculated by equation (2.2).

$$C = f\lambda$$
$$\lambda = \frac{C}{f}$$
$$\frac{\lambda}{4} = \frac{C}{4f}$$

TABLE 4.2   COORDINATES OF UNIT
SPHERE

20-Point Array		
x	y	z
0	0.93	0.36
0	0.93	−0.36
0.58	0.58	0.58
0.58	0.58	−0.58
0.93	0.36	0
0.36	0	0.93
0.36	0	−0.93
0.93	−0.36	0
0.58	−0.58	0.58
0.58	−0.58	−0.58
0	−0.93	0.36
0	−0.93	−0.36
−0.58	−0.58	0.58
−0.58	−0.58	−0.58
−0.93	−0.36	0
−0.36	0	0.93
−0.36	0	−0.93
−0.93	0.36	0
−0.58	0.58	0.58
−0.58	0.58	−0.58

Averaging constant = 13 dB

If $f = 100$ Hz

$$\frac{\lambda}{4} = \frac{1128}{4 \times 100} = 2.82 \text{ ft}$$

This means that if measurements are to be made at a frequency of 100 Hz, the microphone should be at least 2.82 ft away from the walls of the anechoic room.

4.6.3   12-Point Array.   Less accurate results are obtained by using a 12-point array. The coordinates of this array are given in Table 4.3.

4.6.4   8-Point Array.   Still less accuracy is obtained with an 8-point array. The coordinates are shown in Table 4.4.

TABLE 4.3 COORDINATES OF UNIT SPHERE

12-Point Array		
x	y	z
0	0.89	0.45
0.53	0.72	−0.45
0.85	0.28	0.45
0.85	−0.28	−0.45
0.53	−0.72	0.45
0	−0.89	−0.45
−0.53	−0.72	0.45
−0.85	−0.28	−0.45
−0.85	0.28	0.45
−0.53	0.72	−0.45
0	0	1
0	0	−1

Averaging constant = 10.8 dB

TABLE 4.4 COORDINATES OF UNIT SPHERE

8-Point Array		
x	y	z
0	0.82	0.58
0	0.82	−0.58
0.82	0	0.58
0.82	0	−0.58
0	−0.82	0.58
0	−0.82	−0.58
−0.82	0	0.58
−0.82	0	−0.58

Averaging constant = 9 dB

Instead of determining the power passing through each incremental area and adding all the parts, the calculation can be greatly simplified by computing the average sound pressure level in decibels, and then assuming that this is the sound pressure level that would exist if the actual source were replaced by a point source of equal sound power.

4.7 Directivity

Most practical noise sources do not radiate uniformly in all directions. At a fixed distance away from the source, the sound pressure level in any octave band generally shows different levels for different directions. A polar plot of the levels gives the directivity pattern of the source. This pattern is really three-dimensional on a hemispherical surface. Many sources are non-directional in the low frequencies, but become more directional in the higher frequencies.

The directional characteristics are usually given in terms of the directivity index, DI, or the directivity factor, Q.

4.7.1 Directivity Index. The amount by which the sound pressure level in some specified direction, and at a radius r, exceeds the average of the sound pressure levels measured on the surface of a hypothetical sphere, of radius r surrounding the source, is called the directional gain or directivity index. That is,

$$DI_\theta = SPL_\theta - \overline{SPL} \qquad (4.6)$$

where DI_θ = directivity index, in decibels,
   SPL_θ = sound pressure level at radius r and in direction θ, and
   $\overline{SPL}$ = average sound pressure level on surface of a sphere, of radius r, surrounding the source.

4.7.2 Directivity Factor, Q. Directivity factor is defined as the ratio of the sound pressure squared, at some fixed distance and specified direction, to the mean square sound pressure at the same distance averaged over all directions from the source. It is equal to the antilog of one-tenth of the directivity index. That is,

$$Q = \text{antilog} \frac{DI}{10} \qquad (4.7)$$

Directivity information for a machine is given as a function of frequency or frequency bands.

An example of sound power level determination in a free field is given below:

EXAMPLE 4.6 Using an 8-point spherical microphone array and a radius of 15 ft, measure octave band sound pressure levels and calculate the sound power level of a machine.

Table 4.4 shows the x-, y-, and z-coordinates for a unit sphere. Therefore, they must be multiplied by 15 to obtain the corresponding coordinates of a 15-ft sphere, measured from the acoustic center of the source. These dimensions are tabulated below:

Location	x	y	z
1	0	12.3	8.7
2	0	12.3	−8.7
3	12.3	0	8.7
4	12.3	0	−8.7
5	0	−12.3	8.7
6	0	−12.3	−8.7
7	−12.3	0	8.7
8	−12.3	0	−8.7

At each microphone location sound measurements are made in each octave band (or other bandwidth). Let the measured sound pressure levels in the 500-Hz band be as follows:

Location	SPL	Antilog $\dfrac{SPL}{10}$
1	96	39.81×10^8
2	89	7.94×10^8
3	95	31.62×10^8
4	95	31.62×10^8
5	92	15.85×10^8
6	94	25.12×10^8
7	91	12.59×10^8
8	88	6.31×10^8
		170.86×10^8

$$\text{Average} = \frac{170.86 \times 10^8}{8} = 21.36 \times 10^8$$

$$\overline{SPL} = 10 \log 21.36 \times 10^8 = 93.3 \text{ dB}$$

(Since the greatest difference in dB readings in this case is 8, the average could have been found by taking the arithmetic average of the dB readings and adding 1 dB. This would give 93.5 dB, which is close enough.)

In a free field the equation for sound power level of a source is

$$PWL = \overline{SPL} + 10 \log 4\pi r^2 - 10.5 + C_1$$

where C_1 = correction for atmospheric temperature and pressure, from equation (3.6). This correction is usually not important.

$$
\begin{aligned}
PWL &= \overline{SPL} + 10 \log 4\pi + 10 \log r^2 - 10.5 \\
&= \overline{SPL} + 10 \log 12.57 + 10 \log r^2 - 10.5 \\
&= \overline{SPL} + 10.99 + 20 \log r - 10.5 \\
&= \overline{SPL} + 20 \log r + 0.5 \quad\quad\quad\quad\quad\quad (4.8)
\end{aligned}
$$

where PWL = sound power level re 10^{-12} W,

     $\overline{SPL}$ = average sound pressure level re 2.0×10^{-5} N/m², and

     r = radius, in feet, measured from the acoustic center of the source.

By substituting the values of $\overline{SPL}$ and r in equation (4.8),

$$
\begin{aligned}
PWL &= 93.3 + 20 \log 15 + 0.5 \\
&= 93.3 + 20 \,(1.176) + 0.5 \\
&= 93.3 + 23.5 + 0.5 \\
&= 117.3 \text{ dB re } 10^{-12} \text{ W}
\end{aligned}
$$

This calculation must be repeated for each frequency band of interest. In most cases the frequency bands are either octave or one-third octave.

4.7.3 Total Sound Power Level. The total sound power level can be found by adding the levels in all octave bands by the method shown in Example 2.2.

4.7.4 Nondirectional Source. When free-field, nondirectional radiation exists, a single measurement of sound pressure level at a given frequency, and at a known radius, is all that is needed to determine the sound power radiated by the source at that frequency. This is true since the sound pressure level is the same at all points that are equidistant from the source, and $\overline{SPL} = SPL_\theta$.

EXAMPLE 4.7 Calculate the directivity index and directivity factor for each of the eight microphone locations in Example 4.6.

$$\overline{SPL} = 93.3 \text{ dB}$$

$$DI = SPL_\theta - 93.3$$

Location	SPL	DI	$\dfrac{DI}{10}$	Q
1	96	+2.7	+0.27	1.86
2	89	−4.3	−0.43	0.37
3	95	+1.7	+0.17	1.48
4	95	+1.7	+0.17	1.48
5	92	−1.3	−0.13	0.74
6	94	+0.7	+0.07	1.17
7	91	−2.3	−0.23	0.59
8	88	−5.3	−0.53	0.30

It can be seen that if the source were nondirectional, the directivity index at all microphone locations would be zero, and the directivity factor would be unity.

EXAMPLE 4.8   Calculate the sound power, in watts, for the 500-Hz octave band in Example 4.6.

$$PWL = 10 \log \frac{W}{10^{-12}} \tag{2.5}$$

$$\frac{PWL}{10} = \log \frac{W}{10^{-12}}$$

$$\frac{W}{10^{-12}} = \text{antilog} \frac{PWL}{10}$$

$$W = 10^{-12} \text{ antilog} \frac{PWL}{10}$$

Since the sound power level in the 500-Hz band in Example 4.6 was found to be 117.3 dB.

$$W = 10^{-12} \text{ antilog} \frac{117.3}{10}$$

$$= 10^{-12} \text{ antilog } 11.73$$

$$= 10^{-12} \times 0.537 \times 10^{12}$$

$$= 0.537 \text{ W}$$

4.7.5 Calculation of SPL from PWL.   When the sound power level and the directivity factor of a directional source are known, the sound pressure level in any direction can be calculated.

In equation (4.6) it is obvious that the sound pressure level in any particular direction is greater or less than the average sound pressure level by an amount equal to the directivity index. Solving equation (4.8) for *SPL*,

$$\overline{SPL} = PWL - 20 \log r - 0.5 \tag{4.9}$$

From equation 4.6

$$\overline{SPL} = SPL_\theta - DI$$

Therefore,

$$SPL_\theta - DI = PWL - 20 \log r - 0.5$$

$$SPL_\theta = PWL - 20 \log r + DI - 0.5$$

Also,

$$Q = \text{antilog} \frac{DI}{10} \tag{4.7}$$

$$\log Q = \frac{DI}{10}$$

$$DI = 10 \log Q$$

Therefore,

$$SPL_\theta = PWL - 20 \log r + 10 \log Q - 0.5 \tag{4.10}$$

EXAMPLE 4.9 A sound source has a power level of 117.3 dB. Calculate the sound pressure level at an angle of 20° from its axis and a distance of 15 ft, if the directivity factor in that direction is 1.48.

$$SPL_\theta = PWL - 20 \log r + 10 \log Q - 0.5$$

$$= 117.3 - 20 \log 15 + 10 \log 1.48 - 0.5$$

$$= 117.3 - 20 \,(1.176) + 10 \,(0.170) - 0.5$$

$$= 117.3 - 23.5 + 1.7 - 0.5$$

$$= 95. \text{ dB}$$

CHECK.   This checks with location 3 in Example 4.6.

4.8   Sound Power Level in a Free Field above a Reflecting Plane

Unless a machine is small and light in weight it is not feasible to determine its sound power level in a true free field. Most machines are either too heavy to be suspended in a free field, or for some other reason it is impractical to test them there. Furthermore, it is common practice to install industrial machinery on a concrete floor or other hard surface. In such cases the reflecting surface is considered to be part of the source. The determination of sound power level in a free field above a reflecting plane is, therefore, one of practical importance.

The procedure for determining sound power level in a free field above a reflecting plane is similar to that used for free field, except that a hypothetical hemisphere surrounding the sound source is used for sound measurement instead of a spherical surface. The surface of the hemisphere should be in the far field to avoid near-field effects, and, therefore, the radius of the hemisphere should be greater than four times the largest dimension of the source. This requirement is usually difficult to meet, and in practice the near field is considered to extend to twice the largest source dimension, instead of four times, but in no case should the radius be less than 2 ft from the machine under test. Also, as in the case of free field, the surface of the hemisphere should be at least $\lambda/4$ away from the absorbent surfaces of the room, where λ is the wavelength of sound at the lowest frequency of interest.

4.8.1 12-Point Array. The x-, y-, and z-coordinates of a 12-point array are tabulated in Table 4.5.

In the 12-point array it should be noted that four of the microphone locations have z-dimensions at zero; that is, they are in the reflecting plane itself. This means that all other locations in the array, except these four,

TABLE 4.5 COORDINATES OF UNIT HEMISPHERE

12-Point Array		
x	y	z
0	0.93	0.36
0.58	0.58	0.58
0.93	0.36	0 *
0.36	0	0.93
0.93	−0.36	0 *
0.58	−0.58	0.58
0	−0.93	0.36
−0.58	−0.58	0.58
−0.93	−0.36	0 *
−0.36	0	0.93
−0.93	0.36	0 *
−0.58	0.58	0.58

* Subtract 3 dB from the levels measured at these locations.

have equal areas. The four that have a height, z, of zero above the reflecting plane, have only one-half as much area. The other half of each area would be below the floor, in the case of a sphere, but with a hemisphere, the floor cuts it in half. Since only half as much power is radiated through these areas, and since one-half power means 3-dB less, 3 dB must be subtracted from the levels measured at these locations.

This problem can be overcome by tilting the original set of microphone points. Two of the four areas that are cut in half by the floor are now raised above the reflecting plane, and the other two are eliminated. The coordinates of a 10-point array are shown in Table 4.6.

TABLE 4.6   COORDINATES   OF   UNIT HEMISPHERE

10-Point Array		
x	y	z
−0.872	0.357	0.333
−0.333	0.577	0.745
0.127	0.934	0.333
0.745	0.577	0.333
0.667	0	0.745
0.745	−0.577	0.333
0.127	−0.934	0.333
−0.333	−0.577	0.745
−0.872	−0.357	0.333
0	0	1

4.8.2 6-Point Array.   The coordinates of a 6-point array are listed in Table 4.7.

TABLE 4.7   COORDINATES   OF   UNIT HEMISPHERE

6-Point Array		
x	y	z
0	0.89	0.45
0.85	0.28	0.45
0.53	−0.72	0.45
−0.53	−0.72	0.45
−0.85	0.28	0.45
0	0	1

4.8.3 4-Point Array. Table 4.8 gives the coordinates of a 4-point array.

TABLE 4.8 COORDINATES OF UNIT
HEMISPHERE

4-Point Array		
x	y	z
0	0.82	0.58
0.82	0	0.58
0	−0.82	0.58
−0.82	0	0.58

In hemispherical array measurements the acoustic center is assumed to be on the floor, directly underneath the geometric center of the machine.

4.8.4 Wave Interference from Reflecting Plane. An interference effect can occur when sound tests are made above a reflecting plane, and discrete frequency components are present. In Fig. 4.2, whenever the difference between the direct path length from sound source to microphone and that of the reflected wave from source to floor to microphone is an odd multiple of a half wavelength, cancellation can occur.

If the microphone height is changed, then cancellation occurs at some other frequency. If the difference is an even number of half wavelengths, that is, one or more whole wavelengths, the sound pressure level doubles.

Curves *A*, *B*, and *C* of Fig. 4.2 show the results of sound pressure level measurements on an air-operated grinder, with the microphone at three different heights above the floor.

If the microphone is moved up and down in a vertical plane while the measurement is being made, the effect is eliminated, as shown in the bottom curve of Fig. 4.2.

The results of this test emphasize the advantages of selecting different microphone heights when conducting a sound power determination. For example, Table 4.7 lists five of the six microphone locations at 0.45, and Table 4.8, for a 4-point array, shows all microphone locations at a height of 0.58. An interference error at one microphone location would be present at all locations at that same height.

It should be noted that the interference effect occurs in a vertical plane, and it cannot be relieved by moving the microphone horizontally.

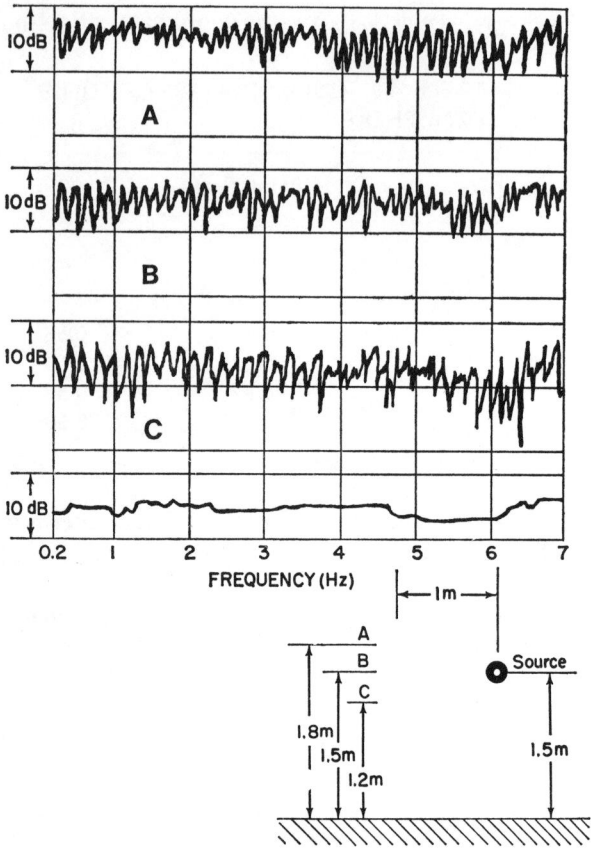

Figure 4.2

In a hemispherical free field above a reflecting plane, the sound power level is

$$PWL = \overline{SPL} + 10 \log 2\pi r^2 - 10.5$$
$$= \overline{SPL} + 10 \log 2\pi + 10 \log r^2 - 10.5$$
$$= \overline{SPL} + 10 \log 6.28 + 10 \log r^2 - 10.5$$
$$= \overline{SPL} + 7.98 + 20 \log r - 10.5$$
$$= \overline{SPL} + 20 \log r - 2.5 \tag{4.11}$$

where PWL = sound power level re 10^{-12} W,

$\overline{SPL}$ = average sound pressure level re 2.0×10^{-5} N/m², and

r = radius, in feet, measured from the acoustic center.

The calculations of sound power level, directivity index, and directivity factor are made for each frequency band of interest, as in the case of spherical radiation, except that equation (4.11) is used instead of equation (4.8).

When a machine is tested in a free field above a reflecting plane, the sound pressure level measurements are 3-dB greater than they would be if the machine were tested in a free field. Although the total sound power output is the same in both cases, it is concentrated in a hemisphere in one case and distributed over a sphere in the other case. This means that for any particular sound power level, the sound pressure level is 3-dB higher in a hemispherical free field above a reflecting plane than it would be in a spherical free field.

For spherical radiation,

$$\overline{SPL} = PWL - 20 \log r - 0.5 \tag{4.9}$$

and for hemispherical radiation,

$$\overline{SPL} = PWL - 20 \log r + 2.5 \tag{4.12}$$

4.9 Sound Power Level in a Reverberant Field

Sound power level can be determined in a reverberant room by the equation:

$$PWL = \overline{SPL} + 10 \log V - 10 \log T - 29.5 \tag{4.13}$$

where PWL = sound power level, in decibels, re 10^{-12} W,

$\overline{SPL}$ = average sound pressure level, in decibels, re 2.0×10^{-5} N/m^2,

V = total air volume of the reverberant room, in cubic feet, with the source in place, and

T = reverberation time, in seconds, for the particular frequency band. This is measured for each frequency band of interest with the sound source in place in the reverberant room.

EXAMPLE 4.10 A room whose dimensions are 25-ft long, 15-ft wide, and 10-ft high has a measured reverberation time of 1.5 sec in the 1000-Hz octave band.

Calculate the sound power level in the 1000-Hz band for a machine if the average sound pressure level in the room is 87 dB in the 1000-Hz octave band.

$$
\begin{aligned}
PWL &= \overline{SPL} + 10 \log V - 10 \log T - 29.5 \\
&= 87 + 10 \log (25 \times 15 \times 10) - 10 \log 1.5 - 29.5 \\
&= 87 + 10 \log 3750 - 10 \log 1.5 - 29.5 \\
&= 87 + 35.7 - 1.8 - 29.5 \\
&= 91.4 \text{ dB re } 10^{-12} \text{ W} \tag{4.13}
\end{aligned}
$$

Figure 4.3

4.10   Determination of Sound Power Level by Means of a Calibrated Sound Source (Courtesy of ILG)

A convenient method for determining sound power level is to compare the unknown machine with a previously calibrated reference sound source, such as that shown in Fig. 4.3. The procedure for doing this is as follows:

1. The average sound pressure level is found for the machine under test, for each frequency band of interest, by measuring sound pressure levels at microphone locations shown in Tables 4.5 to 4.8.
2. The machine under test is replaced by the reference sound source. The measurements are repeated at each microphone location for each frequency band of interest, and the average sound pressure level is calculated.
3. The sound power level of the machine under test is then determined for each frequency band by the equation:

$$PWL_x = \overline{SPL_x} + (PWL_s - \overline{SPL_s}) \qquad (4.14)$$

where PWL_x = band sound power level of the machine under test,

PWL_s = band sound power level of the reference sound source,

$\overline{SPL_x}$ = average sound pressure level measured with the machine under test, and

$\overline{SPL_s}$ = average sound pressure level measured with the reference source.

EXAMPLE 4.11   Calculate the sound power level in the 1000-Hz octave band for a machine if the average sound pressure level measured around the machine in the reverberant room is 82 dB, and the average sound pressure level measured in the same room with the reference source is 66 dB.

Calibration of the reference source gives the following octave band sound power levels:

Octave Band	PWL re 10^{-12} W
63	79.0
125	76.5
250	78.0
500	79.0
1K	79.0
2K	80.0
4K	78.5
8K	77.0

Therefore, the sound power level of the machine in the 1000-Hz octave band is

$$PWL_x = 82 + (79 - 66)$$

$$= 95 \text{ dB re } 10^{-12} \text{ W}$$

4.11   Sound Power Level in a Semireverberant Field

Sound pressure levels in a semireverberant field do not decrease with distance according to the inverse square law as they do in a free field. At the same distance from the source they are somewhat higher in a semireverberant field than they would be in a free field. The relationship between $\overline{SPL}$ and PWL can be shown by the following equation:

$$\overline{SPL} = PWL + 10 \log \left[\frac{1}{4\pi r^2} + \frac{4}{R} \right] + 10.5 \qquad (4.15)$$

where $\overline{SPL}$ = the average sound pressure level, in decibels, re 2.0×10^{-5} N/m^2,

PWL = the sound power level, in decibels, re 10^{-12} W,

r = the radius or distance from the source, in feet, and

R = the room constant, in square feet.

Figure 4.4 shows a plot of relative sound pressure level, in decibels, versus distance from the source, in a semireverberant field for various values of room constant.

The straight line is for a free field, where $R = \infty$, and, therefore, it follows the inverse square law. It can be seen that when the room constant is large, the decrease in sound pressure level with distance from the source remains close to the free-field line longer than when the room constant is small.

If the source is nondirectional, $SPL_\theta = \overline{SPL}$.

In a free field, $R = \infty$, and the equation for sound pressure level is

$$\overline{SPL} = PWL + 10 \log \left[\frac{1}{4\pi r^2}\right] + 10.5$$

$$= PWL - 10 \log 4\pi r^2 + 10.5$$

$$= PWL - 10 \log r^2 - 10 \log 4\pi + 10.5$$

$$= PWL - 20 \log r - 11 + 10.5$$

$$= PWL - 20 \log r - 0.5 \tag{4.9}$$

Note that this equation is identical to equation (4.9).

In the reverberant part of the field, where the curves of Fig. 4.4 are almost

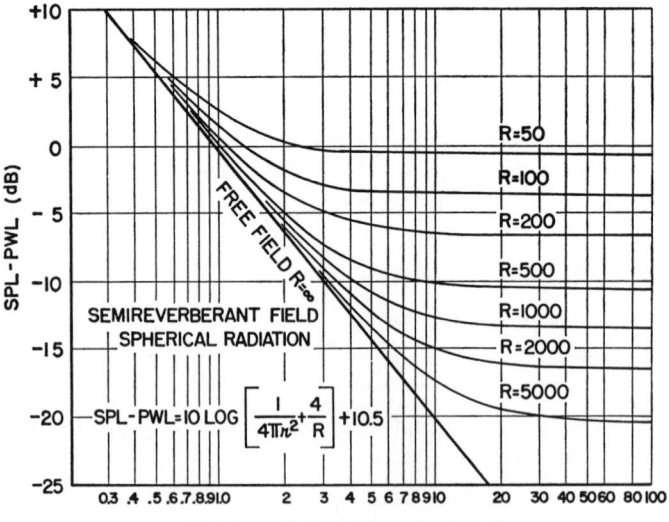

Figure 4.4

parallel to the x-axis, sound pressure level changes very little with distance, and depends only on the room constant, so that

$$\overline{SPL} = PWL + 10 \log \left(\frac{4}{R}\right) + 10.5$$

$$= PWL + 10 \log 4 - 10 \log R + 10.5$$

$$= PWL - 10 \log R + 16.5 \qquad (4.16)$$

EXAMPLE 4.12 A small, nondirectional sound source whose sound power level is 100 dB re 10^{-12} W is located in a semireverberant room whose room constant is 500 ft². Calculate (a) the sound pressure level at a distance of 2 ft from the acoustic center, and (b) the sound pressure level, farther away, in the reverberant part of the far field.

(a)
$$\overline{SPL} = PWL + 10 \log \left[\frac{1}{4\pi r^2} + \frac{4}{R}\right] + 10.5 \qquad (4.15)$$

$$= 100 + 10 \log \left[\frac{1}{16\pi} + \frac{4}{500}\right] + 10.5$$

$$= 100 + 10 \log (0.0199 + 0.008) + 10.5$$

$$= 100 - 15.5 + 10.5$$

$$= 95.0 \text{ dB}$$

(b)
$$\overline{SPL} = PWL - 10 \log R + 16.5 \qquad (4.16)$$

$$= 100 - 10 \log 500 + 16.5$$

$$= 100 - 10 \,(2.7) + 16.5$$

$$= 89.5 \text{ dB}$$

Both these results can be checked on Fig. 4.4.

4.11.1 Sound Power Level Determination. Equation (4.15) can be solved for *PWL* so that it may be used to determine sound power level in a semireverberant room from measured sound pressure levels.

$$PWL = \overline{SPL} - 10 \log \left[\frac{1}{4\pi r^2} + \frac{4}{R}\right] - 10.5 \qquad (4.17)$$

The room constant must be calculated or measured when used in this equation.

4.12 Sound Power Level in a Semireverberant Field above a Reflecting Plane

Industrial machinery is usually installed on a hard reflecting floor or surface. Therefore, considerable importance is placed on sound measurements made in a semireverberant field above a reflecting plane, even though the accuracy of such measurements is not as great as it would be if the measurements were made in a free field or a reverberant field.

Since this case is one of hemispherical radiation, instead of spherical radiation, the equation relating sound pressure level and sound power level is

$$\overline{SPL} = PWL + 10 \log \left[\frac{1}{2\pi r^2} + \frac{4}{R} \right] + 10.5 \tag{4.18}$$

In a free field above a reflecting plane, $R = \infty$ and

$$\overline{SPL} = PWL + 10 \log \left[\frac{1}{2\pi r^2} \right] + 10.5$$

$$= PWL - 10 \log r^2 - 10 \log 2\pi + 10.5$$

$$= PWL - 20 \log r - 8 + 10.5$$

$$= PWL - 20 \log r + 2.5 \tag{4.12}$$

This equation is identical to equation (4.12).

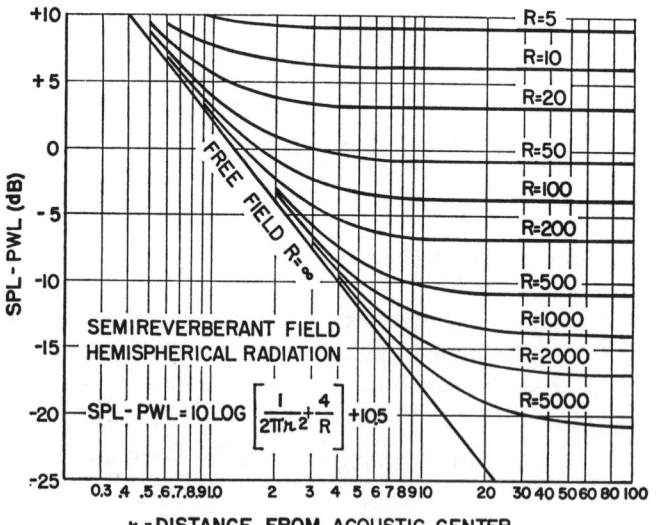

Figure 4.5

Figure 4.5 shows equation (4.18) in the form of curves. It is similar to Fig. 4.4 except that Fig. 4.4 is for spherical radiation, whereas Fig. 4.5 is for hemispherical radiation above a reflecting plane.

EXAMPLE 4.13   Repeat Example 4.12 except assume that the source is placed on a hard reflecting floor.

(a)
$$\overline{SPL} = PWL + 10 \log \left[\frac{1}{2\pi r^2} + \frac{4}{R} \right] + 10.5 \tag{4.18}$$

$$= 100 + 10 \log \left[\frac{1}{8\pi} + \frac{4}{500} \right] + 10.5$$

$$= 100 + 10 \log (0.0398 + 0.008) + 10.5$$

$$= 100 - 13.2 + 10.5$$

$$= 97.3 \text{ dB}$$

NOTE.   The difference in *SPL* between Examples (4.13) and (4.12) is 2.3 dB. If R had been ∞, the difference would have been 3.0 dB.

(b)
$$\overline{SPL} = PWL - 10 \log R + 16.5 \tag{4.16}$$

$$= 100 - 10 \log 500 + 16.5$$

$$= 100 - 10 (2.7) + 16.5$$

$$= 89.5 \text{ dB}$$

4.12.1   *Sound Power Level Determination.*   By solving equation (4.18) for *PWL*,

$$PWL = \overline{SPL} - 10 \log \left[\frac{1}{2\pi r^2} + \frac{4}{R} \right] - 10.5 \tag{4.19}$$

From this equation the sound power level of a machine can be determined in a semireverberant field above a reflecting plane. This is a case that has great practical importance. Room constants for the various frequency bands of interest are either calculated by estimating the absorption in the room and substituting in equation (4.4),

$$R = \frac{S_t \bar{\alpha}}{1 - \bar{\alpha}} \tag{4.4}$$

or by measuring reverberation time and using equation (4.5)

$$R = \frac{S}{(TS/0.049V) - 1} \tag{4.5}$$

Room constant can also be determined by using a nondirectional calibrated sound source.

4.12.2 Determination of Room Constant by Means of a Calibrated Non-directional Sound Source. In a semireverberant field,

$$PWL = \overline{SPL_R} - 10 \log \left[\frac{1}{2\pi r^2} + \frac{4}{R} \right] - 10.5 \qquad (4.19)$$

where $\overline{SPL_R}$ = the sound pressure level in the semireverberant field.

In a free field, sound power level and sound pressure level are related by

$$PWL = \overline{SPL_F} - 10 \log \left[\frac{1}{2\pi r^2} \right] - 10.5$$

where $\overline{SPL_F}$ = the sound pressure level in the free field.

If the sound power level is assumed to remain constant,

$$\overline{SPL_R} - 10 \log \left[\frac{1}{2\pi r^2} + \frac{4}{R} \right] - 10.5 = \overline{SPL_F} - 10 \log \left[\frac{1}{2\pi r^2} \right] - 10.5$$

The difference between these two equations is

$$\overline{SPL_R} - \overline{SPL_F} = 10 \log \left[\frac{1}{2\pi r^2} + \frac{4}{R} \right] - 10 \log \left[\frac{1}{2\pi r^2} \right]$$

$$\frac{\overline{SPL_R} - \overline{SPL_F}}{10} = \log \left[\frac{(1/2\pi r^2) + 4/R}{1/2\pi r^2} \right] = \log \left[1 + \frac{8\pi r^2}{R} \right]$$

$$\text{antilog} \left[\frac{\overline{SPL_R} - \overline{SPL_F}}{10} \right] = 1 + \frac{8\pi r^2}{R}$$

$$\text{antilog} \left[\frac{\overline{SPL_R} - \overline{SPL_F}}{10} \right] - 1 = \frac{8\pi r^2}{R}$$

Solving for the room constant,

$$R = \frac{8\pi r^2}{\text{antilog} \left[\dfrac{\overline{SPL_R} - \overline{SPL_F}}{10} \right] - 1} \qquad (4.20)$$

where R = the room constant, in square feet,

$\quad r$ = the radius of the test hemisphere, in feet,

$\overline{SPL_R}$ = the average octave band sound pressure level in the room under test, in decibels, and

$\overline{SPL_F}$ = the average octave band sound pressure level in a free field above a reflecting plane.

The source is placed on the floor of the room to be tested, and octave band sound pressure levels are measured at some convenient radius.

Octave band sound pressure levels are then measured at that same radius in a free field above a reflecting plane. From these measurements, R can be found for each octave band of interest.

When octave band sound power levels of the reference source are known, from some previous calibration, the free-field sound pressure levels can be calculated from equation (4.12),

$$\overline{SPL} = PWL - 20 \log r + 2.5 \qquad (4.12)$$

instead of measuring them to obtain $\overline{SPL_F}$.

EXAMPLE 4.14   Calculate the room constant in the 1000-Hz octave band by using a calibrated sound source if the average sound pressure level in the room, at a radius of 15 ft, is found to be 65 dB.

The source calibration states that in the 1000-Hz octave its sound power level is 79 dB re 10^{-12} W.

From equation (4.12) the free-field sound pressure level, $\overline{SPL_F}$, is found to be

$$\overline{SPL_F} = PWL - 20 \log r + 2.5$$

$$= 79 - 20 \log 15 + 2.5$$

$$= 79 - 23.5 + 2.5$$

$$= 58. \text{ dB}$$

Therefore, from equation (4.20)

$$R = \frac{8\pi \times 15^2}{\text{antilog} \left[\dfrac{65 - 58}{10} \right] - 1}$$

$$= \frac{1800\pi}{\text{antilog } 0.71 - 1}$$

$$= \frac{1800\pi}{5.01 - 1} = 1410. \text{ ft}^2$$

4.13   Determination of Sound Power Level by Means of Sound Pressure Level Measurements at Various Distances

Sound power level can be determined in certain cases by solving two simultaneous equations, eliminating room constant, and using, instead, sound pressure level measurements made at different distances from the sound source.

Measurements are made in a suitable hemispherical array at two different radii. Since the sound power level is the same in both sets of measurements, and the room constant is assumed to remain the same, the resulting two equations can be solved to eliminate the room constant. It must be recognized that the assumption that the room constant remains unchanged may not be exactly true, especially in the reverberant part of the sound field.

From equation (4.18),

$$\overline{SPL_1} = PWL + 10 \log \left[\frac{1}{2\pi r_1{}^2} + \frac{4}{R} \right] + 10.5$$

$$\overline{SPL_2} = PWL + 10 \log \left[\frac{1}{2\pi r_2{}^2} + \frac{4}{R} \right] + 10.5$$

where $\overline{SPL_1}$ = average sound pressure level measured at smaller radius, r_1,
    $\overline{SPL_2}$ = average sound pressure level measured at larger radius, r_2,
    PWL = sound power level, in decibels, re 10^{-12} W, and
    R = room constant, in square feet.

Let $a = 1/2\pi r_1{}^2$, $b = 1/2\pi r_2{}^2$, and $D = \overline{SPL_1} - \overline{SPL_2}$.

$$\overline{SPL_1} - PWL - 10.5 = 10 \log \left[a + \frac{4}{R} \right]$$

$$\overline{SPL_2} - PWL - 10.5 = 10 \log \left[b + \frac{4}{R} \right]$$

$$\frac{SPL_1 - PWL - 10.5}{10} = \log \left[a + \frac{4}{R} \right]$$

$$\frac{SPL_2 - PWL - 10.5}{10} = \log \left[b + \frac{4}{R} \right]$$

$$10^{(SPL_1 - PWL - 10.5)/10} = a + \frac{4}{R}$$

$$10^{(SPL_2 - PWL - 10.5)/10} = b + \frac{4}{R}$$

$$10^{(SPL_1 - PWL - 10.5)/10} - a = \frac{4}{R}$$

$$10^{(SPL_2 - PWL - 10.5)/10} - b = \frac{4}{R}$$

$$10^{(SPL_1-PWL-10.5)/10} - a = 10^{(SPL_2-PWL-10.5)/10} - b$$

$$10^{(SPL_1-PWL-10.5)/10} - 10^{(SPL_2-PWL-10.5)/10} = a - b$$

$$[10^{SPL_1/10} \times 10^{-PWL/10} \times 10^{-10.5/10}]$$

$$- [10^{SPL_2/10} \times 10^{-PWL/10} \times 10^{-10.5/10}] = a - b$$

$$10^{-PWL/10} \times 10^{-10.5/10}[10^{SPL_1/10} - 10^{SPL_2/10}] = a - b$$

$$\frac{-PWL}{10} - \frac{10.5}{10} + \log[10^{SPL_1/10} - 10^{SPL_2/10}] = \log(a - b)$$

$$\frac{-PWL}{10} - \frac{10.5}{10} = \log(a - b) - \log[10^{SPL_1/10} - 10^{SPL_2/10}]$$

$$\frac{-PWL}{10} - \frac{10.5}{10} = \log\left[\frac{(a-b)}{10^{SPL_1/10} - 10^{SPL_2/10}}\right]$$

$$\frac{-PWL}{10} - \frac{10.5}{10} = \log\frac{(a - b)}{10^{SPL_2/10}[(10^{SPL_1/10}/10^{SPL_2/10}) - 1]}$$

$$\frac{-PWL}{10} - \frac{10.5}{10} = \log\frac{(a - b)}{10^{SPL_2/10}[10^{(SPL_1-SPL_2)/10} - 1]}$$

Since $D = SPL_1 - SPL_2$

$$\frac{-PWL}{10} - \frac{10.5}{10} = \log\frac{(a - b)}{10^{SPL_2/10}[10^{D/10} - 1]}$$

$$-PWL - 10.5 = 10\log\frac{(a - b)}{10^{SPL_2/10}[10^{D/10} - 1]}$$

$$-PWL - 10.5 = 10\log\frac{(a - b)}{10^{D/10} - 1} - SPL_2$$

$$PWL = SPL_2 - 10\log\frac{(a - b)}{10^{D/10} - 1} - 10.5$$

$$PWL = SPL_1 - 10\log\frac{(a - b)}{10^{D/10} - 1} - D - 10.5 \qquad (4.21)$$

EXAMPLE 4.15   A machine is located in a semireverberant shop area on a concrete floor. The average sound pressure level in the 1000-Hz octave at a radius of 5 ft is found to be 82 dB. At a radius of 10 ft the average sound pressure level is 80 dB. Determine the sound power level in the 1000-Hz octave.

$$D = SPL_1 - SPL_2 = 82 - 80 = 2 \text{ dB}$$

$$a = \frac{1}{2\pi r_1^2} = \frac{1}{50\pi} = 0.00637$$

$$b = \frac{1}{2\pi r_2^2} = \frac{1}{200\pi} = 0.00159$$

$$a - b = 0.00478$$

$$PWL = 82 - 10 \log \frac{(0.00478)}{(10^{0.2} - 1)} - 2 - 10.5$$

$$= 82 - 10 \log 0.00478 + 10 \log (10^{0.2} - 1) - 2 - 10.5$$

$$= 82 - 10 (7.679 - 10) + 10 (9.763 - 10) - 2 - 10.5$$

$$= 82 + 23.2 - 2.4 - 2 - 10.5$$

$$= 90.3 \text{ dB re } 10^{-12} \text{ W}$$

These sound pressure level measurements can be used to determine room constant also.

EXAMPLE 4.16   From the sound pressure level measurements in Example 4.15 calculate the room constant in the 100-Hz octave

$$\overline{SPL} = PWL + 10 \log \left[\frac{1}{2\pi r^2} + \frac{4}{R} \right] + 10.5 \qquad (4.18)$$

where $r = 10$ ft, and

$$SPL = 80.$$

From Example 4.15, $PWL = 90.3$ dB.

$$80 = 90.3 + 10 \log \left[\frac{1}{200\pi} + \frac{4}{R} \right] + 10.5$$

$$80 - 90.3 - 10.5 = 10 \log \left[\frac{1}{200\pi} + \frac{4}{R} \right]$$

$$-20.8 = 10 \log \left[0.00159 + \frac{4}{R} \right]$$

$$\log \left[0.00159 + \frac{4}{R} \right] = -2.08$$

$$\left(0.00159 + \frac{4}{R}\right) = \text{antilog } (-2.08) = \text{antilog } (7.92 - 10)$$

$$\left(0.00159 + \frac{4}{R}\right) = 0.00832$$

$$\frac{4}{R} = 0.00832 - 0.00159 = 0.00673$$

$$R = \frac{4}{0.00673} = 595 \text{ ft}^2$$

4.14 Directivity due to Location in Room

A machine may be directional because of the way it is designed. For example, an air-operated machine can radiate more noise at its exhaust opening than at other locations around the machine. But even if it radiates uniformly in all directions, there are directional effects due to the way it is mounted in a room. It is extremely infrequent that a purely nondirectional source is positioned in a true free field, or in the geometric center of an anechoic chamber.

It has been explained that when a machine is placed on a hard, reflecting surface all the sound that would have been directed downward, in spherical radiation, is reflected upward into the hemisphere above the floor. The directivity effect can be still higher, depending on the location of the machine in the room.

Equation (4.15) can be written

$$\overline{SPL} = PWL + 10 \log \left[\frac{Q}{4\pi r^2} + \frac{4}{R}\right] + 10.5 \qquad (4.22)$$

where Q, the directivity factor, assumes various values, depending on where the machine is placed in the room.

In Fig. 4.6, if the machine or sound source is mounted at A, near the center of a large room, $Q = 1$. When it is placed at B, on a hard, reflecting floor, $Q = 2$. At location C, where the floor and a reflecting wall meet, $Q = 4$. And at location D, in a corner where two reflecting walls meet the floor, the total sound power of the machine is radiated into one-eighth of the sphere. Therefore, in this case, $Q = 8$, in equation (4.22).

EXAMPLE 4.17 Calculate the sound pressure level in the 1000-Hz octave at a distance of 10 ft if the machine in Example 4.15 is placed in a corner of the room.

EFFECT OF LOCATION IN ROOM

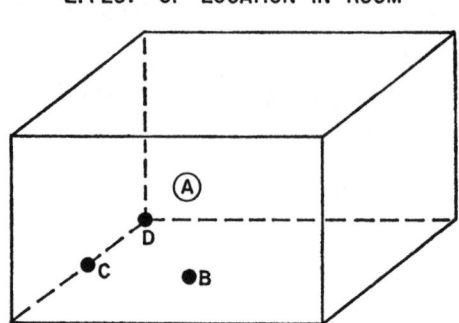

LOCATION	Q
A	1
B	2
C	4
D	8

$$SPL-PWL = 10\,Log\left[\frac{Q}{4\pi r^2} + \frac{4}{R}\right] + 10.5$$

Figure 4.6

For example 4.15 the radiated sound power level is 90.3 W re 10^{-12}, and from Example 4.16 the room constant is 595 ft².

From equation (4.22)

$$\overline{SPL} = PWL + 10\log\left[\frac{Q}{4\pi r^2} + \frac{4}{R}\right] + 10.5 \qquad (4.22)$$

Since $Q = 8$,

$$\overline{SPL} = 90.3 + 10\log\left[\frac{8}{400\pi} + \frac{4}{595}\right] + 10.5$$

$$= 90.3 + 10\log(0.00636 + 0.00672) + 10.5$$

$$= 90.3 + 10\log(0.01307) + 10.5$$

$$= 90.3 + 10(8.1163 - 10) + 10.5$$

$$= 90.3 - 18.8 + 10.5$$

$$= 82.0\ \text{dB}$$

4.15   Effect of Room Constant on Sound Pressure Levels when Measurements are Made Close to Source

In a semireverberant room the sound pressure at any particular distance from a machine, or other sound source, consists of a component radiated directly from the machine plus components reflected from walls, ceiling, and floor.

The curves in Figs. 4.4 and 4.5 show that when the room constant is large, the reflected component is small, and when the room constant is small, the reflected component can be appreciable. The curves also show that when the room constant is large, the room characteristics remain nearer to those of a free field for a greater distance from the source than when the room constant is small.

It is often important to know how room constant affects sound measurements made in typical factory or shop environments.

From equation (4.18) the sound pressure level in a semireverberant field, $\overline{SPL_R}$, is

$$\overline{SPL_R} = PWL + 10 \log \left[\frac{1}{2\pi r^2} + \frac{4}{R} \right] + 10.5 \qquad (4.18)$$

In a free field, $R = \infty$, and

$$\overline{SPL_F} = PWL + 10 \log \frac{1}{2\pi r^2} + 10.5$$

The difference between the sound pressure level measured in the semireverberant field and that in a free field is

$$\overline{SPL_R} - \overline{SPL_F} = 10 \log \left[\frac{1}{2\pi r^2} + \frac{4}{R} \right] - 10 \log \frac{1}{2\pi r^2}$$

$$= 10 \log \left[\frac{(1/2\pi r^2) + 4/R}{1/2\pi r^2} \right]$$

$$= 10 \log \left[1 + \frac{8\pi r^2}{R} \right]$$

Now assume that essentially free-field conditions exist if $\overline{SPL_R} - \overline{SPL_F}$ is not more than 1 dB. This is within the usual measurement accuracy. That is,

$$10 \log \left[1 + \frac{8\pi r^2}{R} \right] = 1$$

$$\log \left[1 + \frac{8\pi r^2}{R} \right] = 0.1$$

$$1 + \frac{8\pi r^2}{R} = 10^{0.1} = 1.259$$

$$\frac{8\pi r^2}{R} = 0.259$$

$$r^2 = \frac{0.259R}{8\pi}$$

$$r = \sqrt{0.0103R} = 0.1015\sqrt{R} \qquad (4.23)$$

This means that if sound measurements are made at a distance of less than 0.1 times the square root of the room constant, the measured sound pressure levels are within 1 dB of what they would be in a free field.

In an average factory the average absorption coefficient is between 0.05 and 0.2. If the room dimensions are 100-ft long, 50-ft wide, and 25-ft high, the total area is 17,500 ft². If the average absorption coefficient is only 0.05, the room constant from equation (4.4) is

$$R = \frac{S_t \bar{\alpha}}{1 - \bar{\alpha}} = \frac{17,500 \times 0.05}{1 - 0.05} = 920 \text{ ft}^2$$

Therefore,

$$r = 0.10\sqrt{R}$$
$$= 0.10\sqrt{920}$$
$$= 0.10 \times 30.4$$
$$= 3.0 \text{ ft}$$

Many sound test codes today require measurements at a distance of 3 ft from the nearest major surface of a machine. This minimizes the effect of reflected noise, and in most cases, such measurements have acceptable accuracy.

4.16 Sound Pressure Level Measurements in Semireverberant Fields Corrected to Approximate Free-Field Conditions

In order to determine whether the noise produced by a machine is in compliance with safety and health regulations, it is necessary to know the sound pressure levels around the machine. Users and purchasers of machinery want to know how much noise the equipment makes when it is installed in their own plant, rather than the levels produced on a manufacturer's test stand, or in a free field.

On the other hand, the machinery manufacturer does not know what the room constant is on the customer's property, and, therefore, must express the machinery noise in terms of sound power level, or in terms of sound pressure level at a fixed distance based on free-field conditions.

In the case of small, movable machinery, this is not much of a problem. But large machines usually cannot be moved from their indoor test stands, and free-field conditions cannot be obtained. For this reason, it is desirable to correct measurements made in the semireverberant room to approximate free-field conditions.

4.16.1 Approximate Free-Field Conditions. In a free field, sound pressure level decreases 6 dB each time the distance from the source is doubled. That is,

$$dB_2 = dB_1 - 20 \log \frac{L_2}{L_1} \tag{4.1}$$

If $L_2 = 2L_1$, and

$$dB_2 = dB_1 - 20 \log 2$$
$$= dB_1 - 20(0.301)$$
$$= dB_1 - 6$$

If a 6-dB decrease in sound pressure level can be obtained as the distance from the source is increased, approximate free-field conditions exist, even though the distance needed to obtain the 6-dB decrease is considerably greater than it would be in a free field.

To illustrate this, first determine the minimum room constant necessary to get a 6-dB decrease in sound pressure level. Next, calculate the sound pressure level with this room constant at a distance of 3 ft from the machine, and compare it to the sound pressure level under free-field conditions.

$$\overline{SPL} = PWL + 10 \log \left[\frac{1}{2\pi r^2} + \frac{4}{R} \right] + 10.5 \tag{4.18}$$

when $R = R$, and
$\quad r = 3$.

$$\overline{SPL_1} = PWL + 10 \log \left[\frac{1}{18\pi} + \frac{4}{R} \right] + 10.5$$

when $R = R$, and
$\quad r = \infty$.

$$\overline{SPL_2} = PWL + 10 \log \left[\frac{4}{R} \right] + 10.5$$

The difference between SPL_1 and SPL_2 is the maximum decrease in sound pressure level that can be obtained with a room constant $= R$, regardless of how far away the microphone is moved.

$$\overline{SPL_1} - \overline{SPL_2} = 10 \log \left[\frac{1}{18\pi} + \frac{4}{R} \right] - 10 \log \left[\frac{4}{R} \right]$$

If $\overline{SPL_1} - \overline{SPL_2} = 6$,

$$10 \log \left[\frac{1}{18\pi} + \frac{4}{R} \right] - 10 \log \left[\frac{4}{R} \right] = 6$$

$$\log \left[\frac{1}{18\pi} + \frac{4}{R} \right] - \log \left[\frac{4}{R} \right] = 0.6$$

$$\log \left[\frac{(1/18\pi) + 4/R}{4/R} \right] = 0.6$$

$$\frac{R}{72\pi} + 1 = 10^{0.6} = 3.98$$

$$\frac{R}{72\pi} = 2.98$$

$$R = 72\pi[2.98] = 674. \text{ ft}^2$$

This is the minimum room constant required to obtain a 6-dB decrease in sound pressure level, independent of distance.

The sound pressure level at a distance $r = 3$ ft from the machine in a semireverberant room with a room constant $R = 674.$ ft^2 is found by equation (4.18).

$$\overline{SPL_R} = PWL + 10 \log \left[\frac{1}{2\pi r^2} + \frac{4}{R} \right] + 10.5$$

$$= PWL + 10 \log \left[\frac{1}{18\pi} + \frac{4}{674} \right] + 10.5$$

The sound pressure level at $r = 3$ ft in a free field can be found by the same equation by setting $R = \infty$.

$$\overline{SPL_F} = PWL + 10 \log \left[\frac{1}{18\pi} \right] + 10.5$$

Therefore,

$$\overline{SPL_R} - \overline{SPL_F} = 10 \log \left[\frac{1}{18\pi} + \frac{4}{674} \right] - 10 \log \left[\frac{1}{18\pi} \right]$$

$$= 10 \log \left[\frac{(1/18\pi) + 4/674}{1/18\pi} \right]$$

$$= 10 \log \left[1 + \frac{72\pi}{674} \right]$$

$$= 10 \log [1 + 0.336]$$

$$= 10[0.1258]$$

$$= 1.25 \text{ dB}$$

That is, if the machine can be considered to be a point source, and if the maximum decrease in sound pressure level with distance from the machine is 6 dB, then the measured sound pressure level 3 ft from the machine is 1.25 dB greater than it would be in a free field.

4.16.2 When 6-dB Decrease cannot be Obtained. Machinery is often installed in locations where a 6-dB decrease cannot be obtained in any direction around the machine. In these cases, sound measurements can be corrected to approximate free-field conditions.

Consider first that the machine can be considered to be a point source. In a semireverberant field above a reflecting plane,

$$\overline{SPL} = PWL + 10 \log \left[\frac{1}{2\pi r^2} + \frac{4}{R}\right] + 10.5 \tag{4.18}$$

For any particular room constant, R, the difference between the sound pressure level at a distance of 3 ft and that at a distance of r ft is

$$\overline{SPL_3} - \overline{SPL_r} = 10 \log \left[\frac{1}{18\pi} + \frac{4}{R}\right] - 10 \log \left[\frac{1}{2\pi r^2} + \frac{4}{R}\right]$$

$$= 10 \log \left[\frac{(1/18\pi) + 4/R}{(1/2\pi r^2) + 4/R}\right] \tag{4.24}$$

When measurements are made at a distance of 3 ft from a machine, the difference between the sound pressure level in a location where the room constant is R and the sound pressure level in a free field is

$$\overline{SPL_R} - \overline{SPL_F} = 10 \log \left[\frac{1}{18\pi} + \frac{4}{R}\right] - \log \left[\frac{1}{18\pi}\right]$$

$$= 10 \log \left[\frac{(1/18\pi) + 4/R}{1/18\pi}\right]$$

$$= 10 \log \left[1 + \frac{72\pi}{R}\right] \tag{4.25}$$

A family of curves can be plotted from equation (4.24) by assigning values to the room constant, R, and for each value of R, letting r equal various distances from 3 ft to about 30 ft.

Equation (4.25) shows the correction to be subtracted from semireverberant field measurements to obtain approximate free-field sound pressure levels. The corrections are marked on the curves (Fig. 4.7) in place of R.

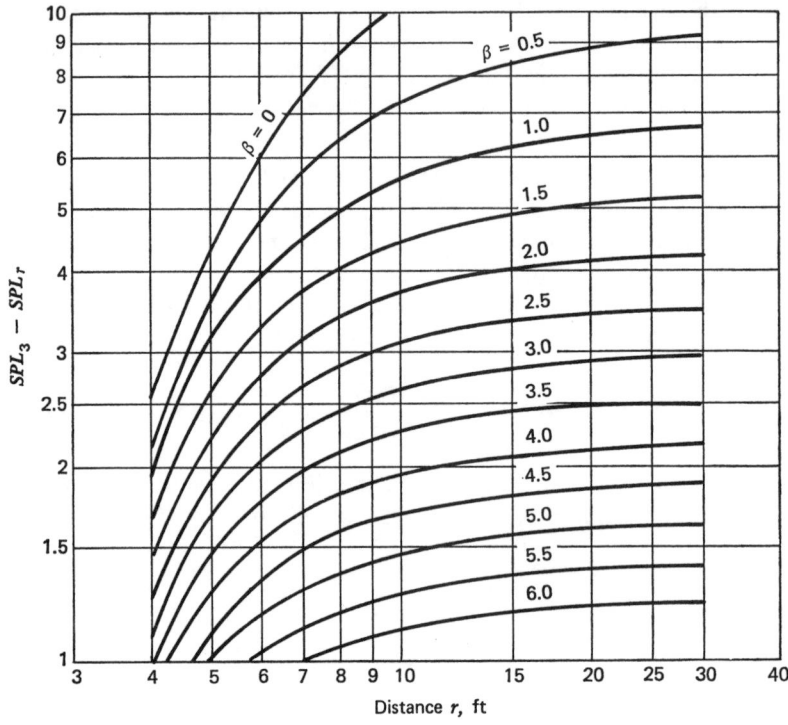

Figure 4.7

EXAMPLE 4.18   A small machine is sound tested in a semireverberant room. At a distance of 3 ft from the machine the sound pressure level in the 500-Hz octave is 92 dB. The microphone is moved away from the machine, and at a distance of 15 ft from the surface of the machine the sound pressure level in the 500-Hz octave is found to be 88 dB. Calculate the approximate free-field sound pressure level.

(a) $SPL_3 = 92$ dBA

(b) $SPL_{15} = 88$ dBA   ($r = 15$ ft)

(c) $SPL_3 - SPL_{15} = 4$ dBA

(d) From Fig. 4.7 the correction is found to be 2.0 dBA.

(e) The approximate free-field sound pressure level in the 500-Hz octave is $92 - 2 = 90$ dBA.

4.17 Sound Power Level Determination—Large Machinery

It is often difficult to determine the sound power level of large machinery. As stated previously, the microphone should be placed at a radius not less than twice the largest dimension of the machine in order to avoid near-field effects. This requirement is usually impossible to obtain in practice, and some compromise must be made.

If the sound power output of a machine is distributed uniformly over an area of S ft², sound power level, sound pressure level, and area are related by the equation

$$PWL = SPL + 10 \log S - 10.5 \qquad (4.26)$$

Today, many sound test codes require that measurements be made at a distance of 3 ft from the nearest major surface of the machine, usually at the ends and at the centers of the sides of the machine itself, the driver, and gears when they are part of the equipment train.

Sound power level can be determined with greater accuracy by increasing the number of measurement locations, as shown on Fig. 4.8. The procedure

SOUND POWER MEASUREMENT
LARGE MACHINERY

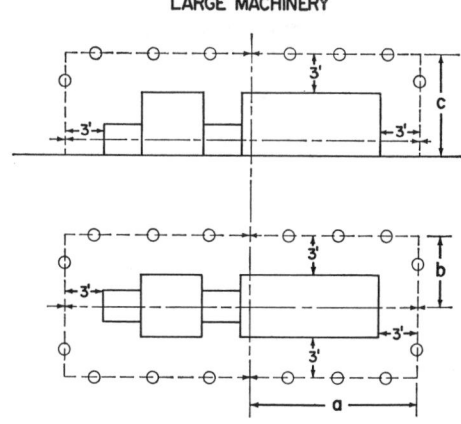

Figure 4.8

is as follows:

1. At each location measure the octave band sound pressure levels.
2. Calculate the average sound pressure level in each octave band as explained in Section 2.10.
3. Assume that the sound power is distributed over the lateral area of one-half an elliptical cylinder, as shown in Fig. 4.9. The length of the cylinder

SOUND POWER MEASUREMENT
LARGE MACHINERY
AREA OF EQUIVALENT HEMISPHERE

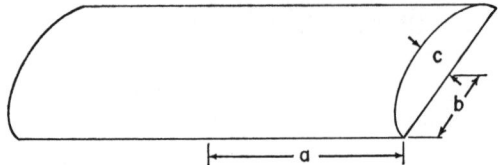

I. LATERAL AREA OF CYLINDER = *(Perimeter of end) x (Length)*

2. PERIMETER OF ELIPTICAL END= π *(b+c)*

3. LATERAL AREA OF CYLINDER= $2\pi a$ *(b+c)*

4. LATERAL AREA OF 1/2 CYLINDER= πa *(b+c)*

5. AREA OF EQUIVALENT HEMISPHERE = πa *(b+c)*

6. $\qquad 2\pi r^2 = \pi a\ (b+c)$

$\qquad\quad 2r^2 = a(b+c)$

$\qquad\quad r^2 = \dfrac{a(b+c)}{2}$

$\qquad\quad r = \left[\dfrac{a(b+c)}{2}\right]^{1/2}$

I. CALCULATE AVERAGE SOUND PRESSURE LEVEL.

$\qquad\qquad \overline{SPL}$

2. CALCULATE RADIUS OF EQUIVALENT HEMISPHERE.

$\qquad\qquad r_s = \left[\dfrac{a(b+c)}{2}\right]^{1/2}$

3. CALCULATE SOUND POWER LEVEL.

$\qquad PWL = \overline{SPL} + 20\ log r_s - 2.5\ re\ 10^{-12}\ watt$

Figure 4.9a–c

is equal to the length of the entire machine plus 6 ft. The width of the cylinder is equal to the width of the machine plus 6 ft, and the height of the cylinder is equal to the height of the machine plus 3 ft.

The lateral area of this half-cylinder is the product of its length and the perimeter of its end, or approximately $\pi a(b + c)$.

4. The radius of a hemisphere with the same area as the cylinder can be found as follows:

$$2\pi r^2 = \pi a(b + c)$$

from which

$$r = \left[\frac{a(b + c)}{2}\right]^{1/2} \tag{4.27}$$

5. Calculate the sound power level in each octave band by means of equation (4.11).

$$PWL = \overline{SPL} + 20 \log r - 2.5 \tag{4.11}$$

where r = the radius of the equivalent hemisphere.

EXAMPLE 4.19 An air compressor 20-ft long, 12-ft wide, and 7-ft high is sound-tested under essentially free-field conditions. The average sound pressure level measured around the machine at a distance of 3 ft is 87 dB in the 1000-Hz octave.

Calculate the sound power level in the 1000-Hz octave.

(a) Radius of equivalent hemisphere

$$r = \left[\frac{a(b + c)}{2} \right]^{1/2}$$

$$a = 13, \; b = 9, \; c = 10$$

$$r = \left[\frac{13(9 + 10)}{2} \right]^{1/2}$$

$$= \sqrt{123.5} = 11.1 \; \text{ft}$$

(b)

$$PWL = \overline{SPL} + 20 \log r - 2.5$$

$$= 87 + 20 \log 11.1 - 2.5$$

$$= 87 + 20.9 - 2.5$$

$$= 105.4 \; \text{dB}$$

 4.18 Correction Factors for Sound Power Level Measurements on Large Machinery Tested Indoors (Two-surface Method)

As shown in Section 4.16, approximate free-field conditions can be assumed to exist if a 6-dB decrease in sound pressure level can be obtained as the distance from the source is increased, even though the distance needed to obtain the 6-dB decrease is considerably greater than it would be in a free field.

When large machinery must be tested in semireverberant locations, corrections can be made in each calculated band sound power level to determine the approximate free-field sound power levels. A method for doing this is as follows:

In Fig. 4.10 octave band sound pressure level measurements are made at predetermined locations on the surface of a rectangular parallelepiped, which

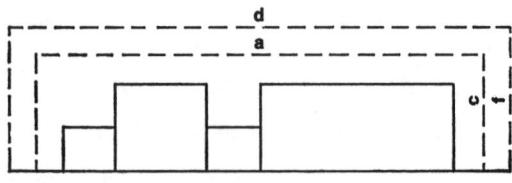

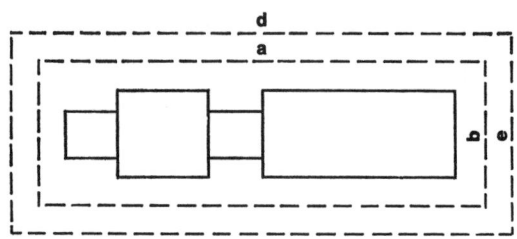

Figure 4.10

has dimensions a, b, and c, as shown. The length, a, is equal to the length of the machine plus 6 ft, the width, b, is equal to the width of the machine plus 6 ft, and the height, c, equals the height of the machine plus 3 ft. That is, sound pressure levels are measured at a distance of 3 ft from the nearest major surfaces.

Next, octave band sound pressure levels are measured at other locations on the surface of a second parallelepiped whose dimensions are d, e, and f, somewhat greater than those of the first set.

Then, from the average sound pressure levels on the two surfaces, and the areas of the surfaces, a correction factor can be calculated. This can be applied to the calculated sound power level to obtain the approximate free-field sound power level.

In Fig. 4.10 let the area of the first parallelepiped be S_1, and the area of the second, larger parallelepiped be S_2.
Then

$$S_1 = ab + 2ac + 2bc \tag{4.28}$$

and

$$S_2 = de + 2df + 2ef \tag{4.29}$$

Sound power level in a semireverberant field above a reflecting plane is

$$PWL = \overline{SPL} - 10 \log \left[\frac{1}{S} + \frac{4}{R} \right] - 10.5 \tag{4.30}$$

where $S =$ the area into which the sound power radiates.

Since the same sound power is radiated into each of the two enclosing areas,

$$PWL = \overline{SPL_1} - 10 \log \left[\frac{1}{S_1} + \frac{4}{R}\right] - 10.5 \qquad (4.31)$$

and

$$PWL = \overline{SPL_2} - 10 \log \left[\frac{1}{S_2} + \frac{4}{R}\right] - 10.5 \qquad (4.32)$$

where PWL = the octave band sound power level, in decibels, re 10^{-12} W,
$\overline{SPL_1}$ = the average octave band sound pressure level on area S_1, in decibels,
$\overline{SPL_2}$ = the average octave band sound pressure level on area S_2, in decibels,
S_1 = the area of smaller parallelepiped, in square feet,
S_2 = the area of larger parallelepiped, in square feet, and
R = the room constant, in square feet.

By solving equations (4.31) and (4.32) for S_1 and S_2, and subtracting,

$$\overline{SPL_1} - \overline{SPL_2} = 10 \log \left[\frac{1}{S_1} + \frac{4}{R}\right] - 10 \log \left[\frac{1}{S_2} + \frac{4}{R}\right]$$

$$= 10 \log \left[\frac{(1/S_1) + 4/R}{(1/S_2) + 4/R}\right]$$

Let $\overline{SPL_1} - \overline{SPL_2} = \Delta Lp$. Then

$$\log \left[\frac{(1/S_1) + 4/R}{(1/S_2) + 4/R}\right] = \frac{\Delta Lp}{10}$$

Or

$$\frac{(1/S_1) + 4/R}{(1/S_2) + 4/R} = 10^{\Delta Lp/10}$$

$$\frac{1}{S_1} + \frac{4}{R} = 10^{\Delta Lp/10}\left[\frac{1}{S_2} + \frac{4}{R}\right]$$

Let

$$10^{\Delta Lp/10} = K \qquad (4.33)$$

Then

$$\frac{1}{S_1} + \frac{4}{R} = K\left[\frac{1}{S_2} + \frac{4}{R}\right]$$

$$= \frac{K}{S_2} + \frac{4K}{R}$$

$$\frac{4}{R} - \frac{4K}{R} = \frac{K}{S_2} - \frac{1}{S_1}$$

$$\frac{4}{R}(1 - K) = \frac{K}{S_2} - \frac{1}{S_1}$$

$$\frac{4}{R} = \frac{1}{1 - K}\left[\frac{K}{S_2} - \frac{1}{S_1}\right] \tag{4.34}$$

Rewriting the form of equation (4.31),

$$PWL = \overline{SPL_1} - 10\log\left[\frac{1}{S_1} + \frac{4}{R}\right] - 10.5$$

$$= \overline{SPL_1} - 10\log\left[\frac{(S_1/S_1) + 4S_1/R}{S_1}\right] - 10.5$$

$$= \overline{SPL_1} - 10\log\left[1 + \frac{4S_1}{R}\right] + 10\log S_1 - 10.5$$

Let

$$C = 10\log\left[1 + \frac{4S_1}{R}\right] \tag{4.35}$$

Then

$$PWL = \overline{SPL_1} + 10\log S_1 - C - 10.5 \tag{4.36}$$

By substituting in equation (4.35) the value of $4/R$ obtained in equation (4.34),

$$C = 10\log\left[1 + \frac{S_1}{1 - K}\left(\frac{K}{S_2} - \frac{1}{S_1}\right)\right]$$

$$= 10\log\left[1 + \frac{1}{1 - K}\left(\frac{KS_1}{S_2} - 1\right)\right]$$

$$= 10\log\left[\frac{1 - K}{1 - K} + \frac{1}{1 - K}\left(\frac{KS_1}{S_2} - 1\right)\right]$$

$$= 10 \log \left[\frac{1 - K}{1 - K} + \frac{(KS_1/S_2) - 1}{1 - K} \right]$$

$$= 10 \log \frac{1}{1 - K} \left[1 - K + \frac{KS_1}{S_2} - 1 \right]$$

$$= 10 \log \frac{1}{1 - K} \left[\frac{KS_1}{S_2} - K \right]$$

$$= 10 \log \frac{K}{1 - K} \left[\frac{S_1}{S_2} - 1 \right] \tag{4.37}$$

where

$$K = 10^{\overline{(SPL_1 - SPL_2)}/10} \tag{4.38}$$

EXAMPLE 4.20 A compressor is driven by a turbine. The length of the equipment is 20 ft, the width is 10 ft, and the height is 7 ft.

The average of the sound pressure levels in the 500-Hz octave, measured around the machine at a distance of 3 ft from the nearest major surface, is 95 dB.

At a distance of 10 ft from the nearest major surface, the average sound pressure level in the 500-Hz octave is 92 dB.

Calculate the sound power level of the turbine-compressor unit in the 500-Hz octave band.

(a)
$$a = 20 + 3 + 3 = 26 \text{ ft}$$
$$b = 10 + 3 + 3 = 16 \text{ ft}$$
$$c = 7 + 3 = 10 \text{ ft}$$
$$d = 20 + 10 + 10 = 40 \text{ ft}$$
$$e = 10 + 10 + 10 = 30 \text{ ft}$$
$$f = 7 + 10 = 17 \text{ ft}$$
$$\overline{SPL_1} = 95 \text{ dB}$$
$$\overline{SPL_2} = 92 \text{ dB}$$

(b)
$$S_1 = ab + 2ac + 2bc$$
$$= 416 + 520 + 320$$
$$= 1256. \text{ ft}^2$$
$$S_2 = de + 2df + 2ef$$
$$= 1200 + 1360 + 1020$$
$$= 3580 \text{ ft}^2$$

(c)
$$K = 10^{(\overline{SPL_1} - \overline{SPL_2})/10} = 10^{(95-92)/10}$$
$$= 10^{0.3} = 2.0$$

(d)
$$\frac{S_1}{S_2} = \frac{1256}{3580} = 0.351$$

(e)
$$C = 10 \log \frac{K}{1-K} \left[\frac{S_1}{S_2} - 1 \right]$$
$$= 10 \log \left[\frac{2}{1-2} \right] [0.351 - 1]$$
$$= 10 \log (-2.)(-0.649)$$
$$= 10 \log 1.298$$
$$= 10(0.113)$$
$$= 1.13 \text{ dB}$$

(f) Therefore, the corrected sound power level in the 500-Hz band is

$$PWL = \overline{SPL_1} + 10 \log S_1 - C - 10.5 \qquad (4.36)$$
$$= 95 + 10 \log 1256 - 1.13 - 10.5$$
$$= 95 + 10\,(3.099) - 1.13 - 10.5$$
$$= 114.4 \text{ dB re } 10^{-12} \text{ W}$$

This calculation must be made for each octave band.

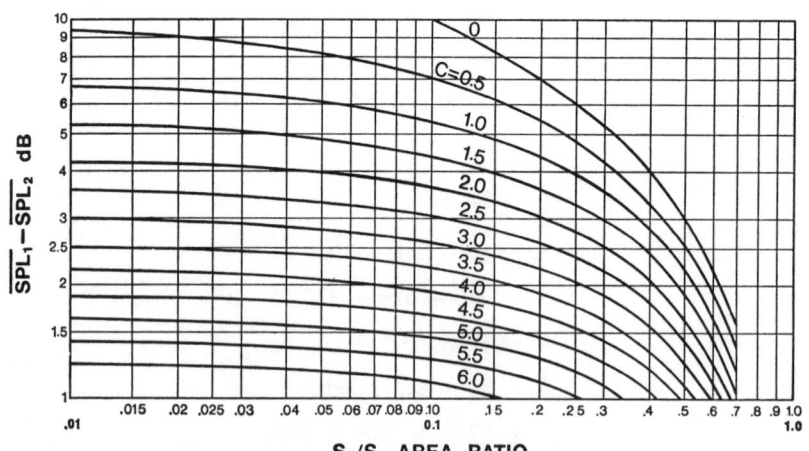

Figure 4.11

Determination of the correction factor, C, can be simplified by plotting equation (4.37) in a series of curves, as shown on Fig. 4.11. The ordinate is plotted in terms of $SPL_1 - SPL_2$, and the abscissa is shown as various values of area ratio S_1/S_2.

For the conditions given in Example 4.20, the point of intersection of the vertical line at $S_1/S_2 = 0.35$ and the horizontal line at $\overline{SPL_1} - \overline{SPL_2} = 3.0$ is at $C = 1.1$ dB.

PART TWO

MACHINERY NOISE SOURCES
AND SOUND CONTROL

CHAPTER 5

MACHINERY NOISE SOURCES

5.1   General

5.1.1   Effect of Horsepower.   The sound produced by a machine is related to the horsepower input to the machine. This does not imply that in the case of an inefficient machine all the excess horsepower goes into noise. Nor does it mean that if the machine had been designed more efficiently it would have been less noisy. However, in practically every case there is a direct relation between horsepower input and generated noise.

Somewhat less obvious is how the noise varies with the horsepower. It is not the same for all types of machinery, and the problem is complicated by the fact that when horsepower input changes, other parameters change also. If it is assumed that the percentage of input horsepower going into noise remains constant, then doubling the horsepower should double the sound power, and this would mean an increase of 3 dB. In most cases this is not true. Doubling the horsepower of a centrifugal compressor or pump results in an increase in sound level of about 4 to 5 dB. This indicates that the machine becomes more efficient in its ability to radiate noise as the horsepower increases.

The effect of horsepower on generated noise for a particular class of compressors, pumps, or other machinery can be obtained experimentally by conducting sound tests under controlled conditions and by developing an empirical equation. With no test data available, an approximate equation for the increase in sound pressure level is

$$\text{Increase in dB} = 17 \log \text{hp ratio} \qquad (5.1)$$

5.1.2   Effect of Rotation Speed.   In general, high-speed machines are noisier than low-speed machines. The increase in noise, as a function of speed, depends on the type of machine, how it is mounted, the ratio of rotating mass to casing mass, how well the machine is aligned, and whether structural resonances are excited.

There is no simple equation relating noise to speed, but for any particular class of machinery the relationship can be determined fairly accurately by conducting sound tests under controlled conditions and by developing an empirical equation.

In general, the overall noise of centrifugal compressors and centrifugal pumps increases in proportion to 20 to 50 times the logarithm of the speed ratio.

5.1.3 Effect of Dynamic Balance. Improving the dynamic balance of rotating machinery often reduces noise levels over the entire spectrum, but the amount of the decrease is relatively small, unless the unbalance is large to start with. Mechanical forces due to unbalance are proportional to the square of the speed, and, therefore, the effects are more pronounced at high speed than at low speed. It is practically impossible, on an analytical basis, to relate sound pressure level change to ounce-inch unbalance change.

When dynamic balancing is necessary, it is always better to apply balance corrections in the planes where the original unbalance occurs; that is, it is best to use multiplane balancing. This is comparatively difficult to do once the machine has been installed, but it can be done easily during assembly, for example, when several impellers are being mounted on a shaft and can be balanced one at a time.

Two-plane balancing is often used when rotating machinery must be balanced in the field, but it should be remembered that in the case of a flexible rotor, a balance correction in only two planes is usually correct at only one speed.

Two-plane vector balancing is done by first measuring the original vibration, and then observing the effect at each end of the machine produced by adding a trial weight at the near end. A trial weight is then added to the far end, and the effect on both ends is measured. From these measurements and the shift of the phase angle, vector calculations are made to determine the amounts and locations of the true balance corrections. The calculations are long and tedious, and sometimes a special computer is used for the purpose.

A somewhat simpler technique is to balance the near end, and then balance the far end. This usually upsets the near-end balance, and it must be worked on again. Since the change in the near end once again affects the far end, it too must be touched up.

The disadvantage of the method is that it means starting and stopping the machine more often than the two-plane vector method. The advantages are that it requires a minimum of equipment, and it almost always works.

Vibration due to unbalance can be the cause of noise radiated from parts of the machine, foundation, and sections of the building in which the machine is installed. In such cases the machine must be dynamically balanced on the job. The procedure is described below:

1. The amplitude of vibration caused by a certain unbalance cannot be determined directly because of the system resonance. This resonance also produces an angular lag between the actual "heavy spot," or place where

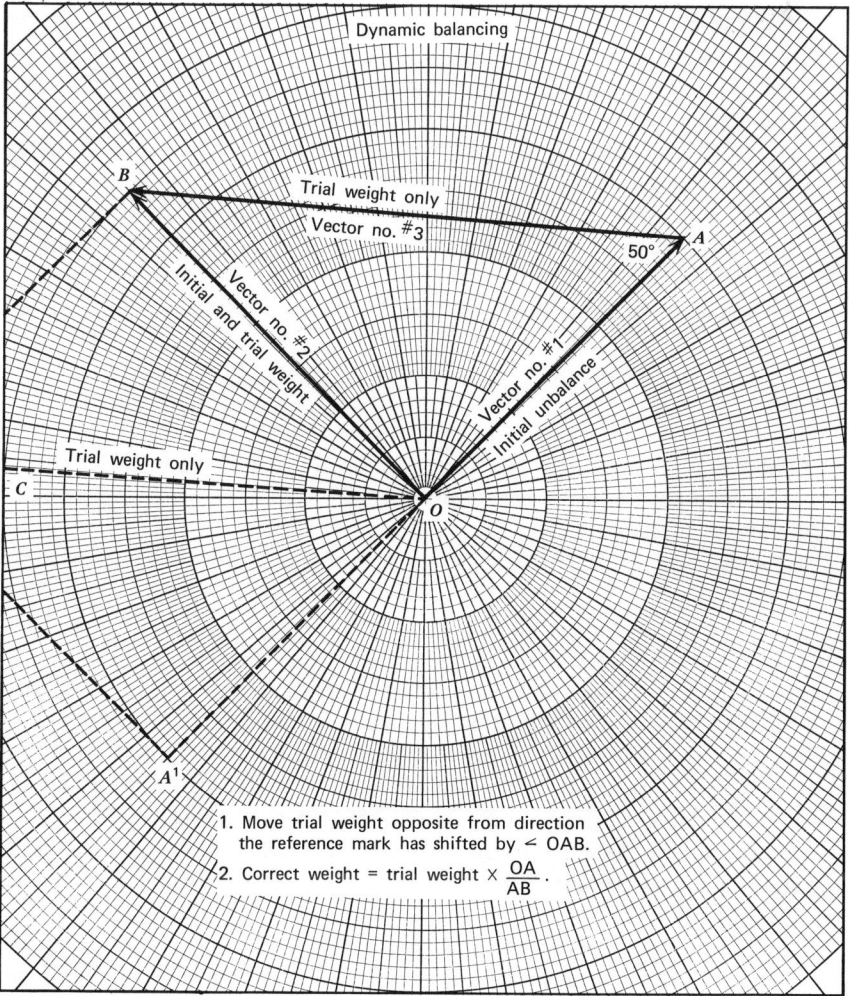

Figure 5.1

weight would have to be removed to obtain balance, and the "high spot," or part that moves out during vibration. The lag is the angle by which the high spot follows the heavy spot in the rotating system.

2. Place a reference mark on the shaft or coupling for phase determination.
3. Measure the vibration double amplitude in peak to peak mils and by means of a strobe light observe the location where the reference mark appears. Figure 5.1 shows the initial vibration double amplitude and direction plotted as OA.
4. Add a trial weight and show the new peak to peak mils vibration and the new reference mark direction at OB. This is the effect of the trial weight added to the initial unbalance.
5. The effect of the trial weight alone is equal to vector OB minus vector OA. This can be obtained as a vector by reversing OA and adding it to OB. The difference, OC, is equal in both magnitude and direction to vector AB. AB represents the effect of the trial weight only.
6. The trial weight should now be moved in a direction opposite to the direction that the reference mark has shifted, by an angle equal to angle AOB.

The correct weight should be calculated as follows:

$$\text{Correct weight} = \text{trial weight} \times \frac{OA}{AB}$$

EXAMPLE 5.1   Assume that the vibration pickup is at 12:00 o'clock and the initial peak to peak vibration is 6 mils with the reference mark appearing at 1:30 o'clock, as shown on Fig. 5.2.

Below the first critical speed, the high spot lags the heavy spot by some angle.

Assume that rotation is clockwise and the actual heavy spot is at 2:00 o'clock, as shown on Fig. 5.2.

Now assume that the trial weight shifts the reference mark to 10:30 o'clock and the new peak to peak vibration amplitude is 7 mils.

Figure 5.2 shows the original condition, and Fig. 5.3 shows the effect of the trial weight. The original heavy spot was clockwise 15° from the reference mark. The trial weight does not change this. That is, there will be a new heavy spot due to the effect of the original heavy spot plus the added weight but the original heavy spot alone will still be 15° clockwise from the reference mark.

If the vibration detector remains in the same location, and the angular distance between the heavy spot and the high spot does not change, then the new heavy spot will still be at 2:00 o'clock as indicated on Fig. 5.3. (That is, the angle of lag is assumed to be unchanged.)

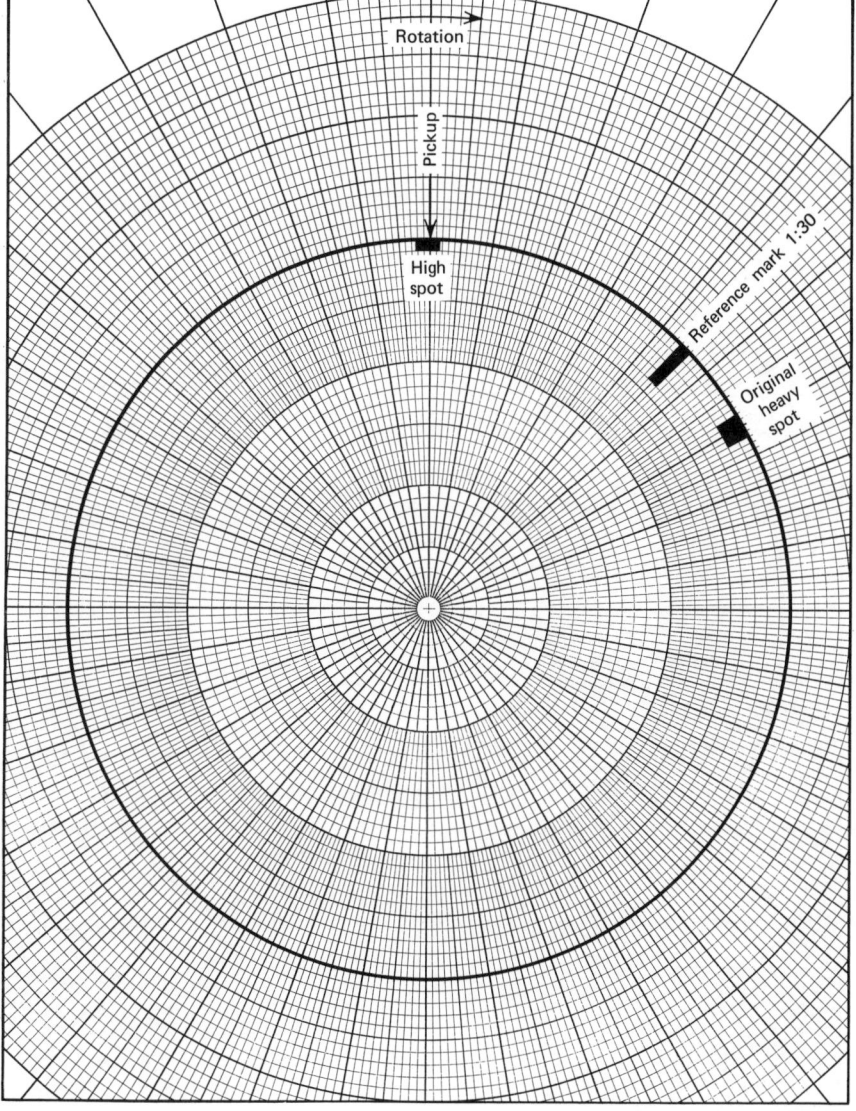

Figure 5.2

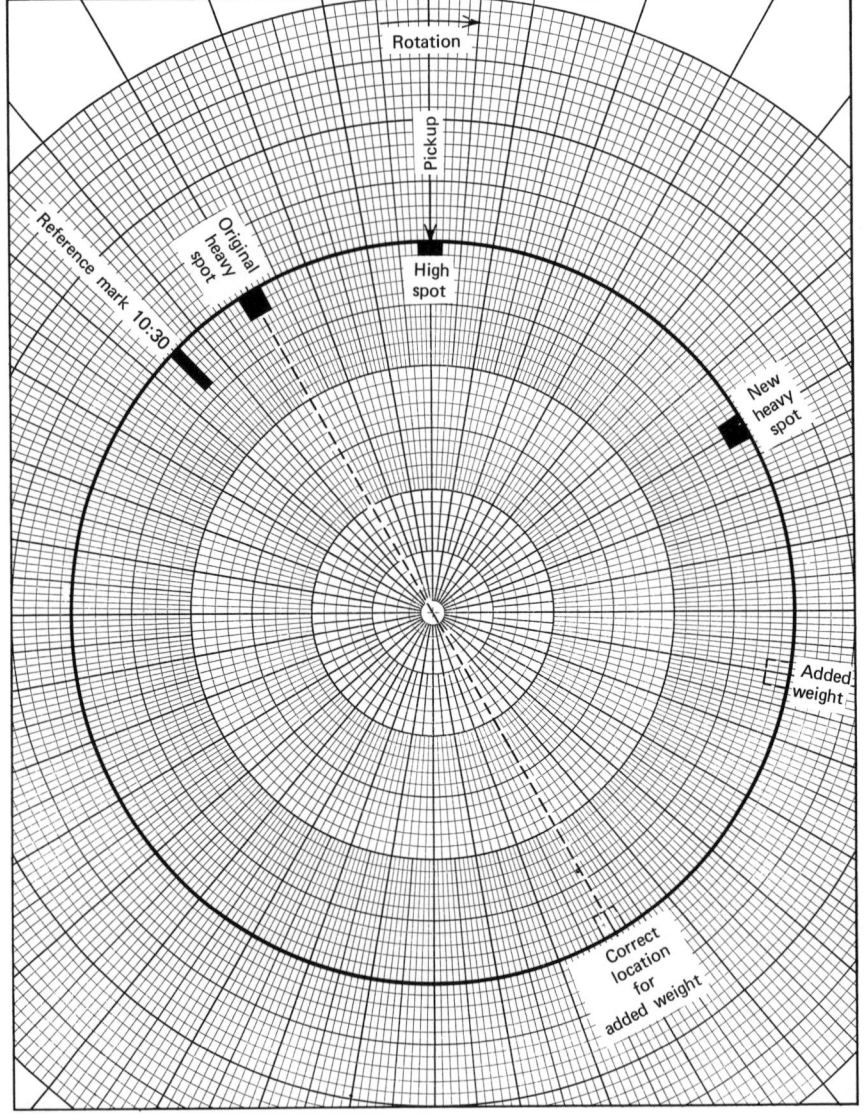

Figure 5.3

Now the new heavy spot is the result of the original heavy spot plus the trial weight. Remembering that the new amplitude is 7 mils and the original amplitude is 6 mils, the location of the added weight can be determined as shown on Fig. 5.4 Here the reference mark is shown at 10:30 o'clock. The original heavy spot is clockwise 15° from the reference mark, and it appears at 30° counterclockwise from the pickup location (see Fig. 5.3). This vector is plotted as OA on Fig. 5.4. The new heavy spot is shown at OB on Fig. 5.4, and it appears at 2:00 o'clock, or 60° clockwise from the pickup. $OB = 7$ mils.

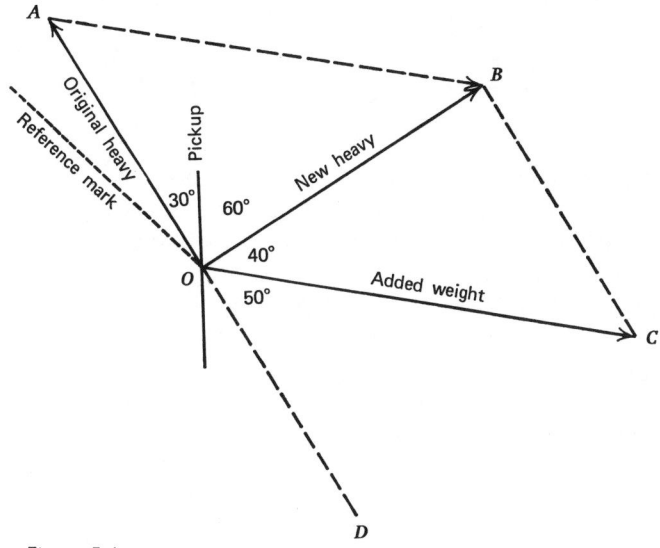

Figure 5.4

OB is the resultant of OA and OC. Therefore, OC must be 100° clockwise from the pickup.

In order to correct the original unbalance, the added weight should be placed clockwise from its present position so that it produces a vector OD, equal in length, but opposite in direction to OA, the original unbalance.

That is, the trial weight should be moved 50° clockwise from its present position, and the correct weight should be found as follows:

$$\text{Correct weight} = \text{trial weight} \times \frac{OA}{OC}$$

5.1.4 Structural Resonances. Structural resonances often are responsible for many components in the sound spectrum, which are not related to any of the obvious forcing frequencies. These resonances can be excited by unbalance, impacting parts, or sliding or rubbing contacts. Not infrequently the vibrating part is internal. To locate and correct it can be a difficult project.

In the design of quiet machinery it is almost mandatory that no major structural resonances be coincident with any strong forcing frequency. Electromechanical shaker tests can usually detect vibration frequencies and mode shapes when any doubt exists concerning their presence.

5.1.5 Impacts.   Airborne sound impinging on a surface produces vibration, and vibrating surfaces radiate airborne sound. The vibration can also travel through a machine as structure-borne sound and be radiated as airborne sound at another more remote point.

Structure-borne sound can also be produced directly by impacting parts that excite resonant vibrations in machine parts. This vibration can result in relatively high levels of airborne sound.

5.1.6 Valves and Metering Orifices.   Control valves are major noise sources in industrial plants. In many instances it is useless to attempt to reduce the noise of a machine unless control valve noise is reduced first.

The three main causes of control valve noise are aerodynamic, cavitation, and mechanical. Aerodynamic noise is the most common and generates the highest levels. It is caused by fluid turbulence and shock waves due to high velocity and mass flow. It has been found that noise increases with velocity when the high velocity is produced by pressure differential across the valve, but that noise increases very little with velocity when the higher velocity is due only to using a gas with a lower molecular weight.

It is sometimes thought that the Strouhal number is a very important factor in predicting noise levels of valves, but many tests indicate no correlation whatever. Strouhal number is given by the following relation:

$$K = \frac{fd}{V}$$

where K = the Strouhal number,
f = the frequency,
d = the pipe diameter, and
v = the velocity.

In most valves the increase in noise is proportional to the seventh or eighth power of the velocity. Actually, pressure-reducing valve noise increases with flow and pressure reduction until the critical pressure ratio is reached. After this, the noise level increases only with an increase in flow. The critical ratio for most gases is slightly less than 2:1.

Acoustic velocity, at Mach 1.0, always produces high noise levels because of turbulence and shock waves, but when the mass flow is large, relatively high noise levels are generated at a Mach number as low as 0.4.

Less noise would be produced if the pressure were reduced in steps, each below the critical ratio. This would be quite expensive, of course.

Cavitation in fluid systems can be a major source of noise. This occurs at flow velocities that are quite critical, but that are difficult to predict because of the influence of geometry, temperature, and pressure.

Cavitation is the result of the collapse of vapor bubbles that have been formed by some unstable static pressure condition in the fluid system. The most typical case is where a flow restriction increases the velocity and decreases the static pressure. When the static pressure falls below the vapor pressure, bubbles form. As they proceed downstream, past the restriction, the velocity decreases and the pressure increases. When pressure increases to a certain point, bubbles collapse to cause pressure fluctuations.

When cavitation is not present in fluid systems, flow noise increases with velocity approximately as

$$60 \log \frac{V_2}{V_1}$$

This means that if the velocity is doubled, the noise increases about 18 dB. When cavitation occurs, the noise can increase about

$$120 \log \frac{V_2}{V_1}$$

Almost all valves cavitate when the flow rate is great enough. Cavitation occurs at a reduced flow rate when the valve is closed partially because of the greater area restriction. Therefore, throttling valves should be located as far away as possible from a machine that is to be operated quietly, or when a sound test is made on it.

It is often quite difficult to reduce the effects of noisy valves. In water, sound is attenuated about 0.00003 dB/ft, so there is very little loss between the noise source and the point of measurement.

Mechanical vibration in valves is caused by turbulence and large mass flows. The resulting noise levels are usually well below that produced by aerodynamic causes, but can be high if certain valve parts are excited at their natural frequencies.

5.1.7 Gears. Gear noise is related to design, manufacturing tolerances, and operation. The rolling and sliding metal to metal contact of surfaces that are neither perfectly smooth nor geometrically correct causes both noise and vibration. Accuracy of tooth form and spacing must be maintained for quiet operation. If the tooth form is not correct, there is a small change in acceleration from one tooth to the next. These speed changes result in both tangential and radial forces, which in turn can cause torsional vibration and excitation of other parts. If the tooth spacing is not correct, there is rough contact on the next tooth. The resulting small impacts cause noise.

Friction is one of the major sources of gear noise, and it probably is the most important one. At the pitch line, the radial force suddenly changes direction from pushing to pulling at a rate equal to the tooth-passing frequency, that is, the number of teeth times revolutions per second. It should be noted that this major source of noise is not dependent on the accuracy of the gear, and even if the gear were 100-percent perfect, noise still would be generated by friction.

In general, gear frequency noise consists of the tooth-passing frequency and a number of higher harmonics plus other components associated with the impacts and structural resonances.

5.1.8 Bearings. Ball bearing noise comes from a number of sources that usually can be traced to surface irregularities on balls or raceways. The cause usually can be found by making a narrow band frequency analysis and comparing the peaks with frequencies calculated from the dimensions of the bearing, the number of rolling elements, and the speed of rotation according to the following equations:

r_1 = radius of inner raceway, in inches,
r_2 = radius of outer raceway, in inches,
r_B = radius of rolling element, in inches,
r_T = radius of train of rolling elements, in inches,
m = number of rolling elements,
n_R = speed of inner raceway, or shaft, in rps,
n_T = speed of train of rolling elements, in rps,
n_B = spin (rotational speed) of rolling elements, in cps,
f_R = fundamental rotational frequency of shaft, in cps,
f_T = fundamental rotational frequency of train, in cps,
f_B = fundamental rotational frequency of rolling element, in cps,
f_1 = frequency due to inner raceway, in cps,
f_2 = frequency due to outer raceway, in cps, and
$r_T = r_1 + r_B.$

(a) The fundamental rotational frequency is generated by any unbalance or eccentricity.

$$f_r = n_R \tag{5.2}$$

(b) Noise produced by rotation of the train of rolling elements due to an irregularity in either a rolling element or the cage.

$$f_T = n_R \cdot \frac{r_1}{r_1 + r_2} \tag{5.3}$$

(c) Noise produced by the spinning of the rolling elements (a rough spot or indentation may strike the inner and outer races alternately).

$$f_B = \frac{r_2}{r_B} \cdot n_R \cdot \frac{r_1}{r_1 + r_2} \tag{5.4}$$

(d) Noise produced by an irregularity on the inner raceway.

$$f_1 = n_R \left(1 - \frac{r_1}{r_1 + r_2} \right) m \tag{5.5}$$

(e) Noise produced by an irregularity on the outer raceway.

$$f_2 = n_R \cdot \frac{r_1}{r_1 + r_2} \cdot m \tag{5.6}$$

Sound tests are usually made to select bearings for quiet operation in cases where bearing noise is likely to be an important factor. These tests are made on an anderometer as the bearing rotates at constant speed. The number of "anderons" is a measure of the radial displacement, with respect to angular displacement, and its units are microinches per radian per second.

In plain bearings, friction is the major source of noise, and eccentricity is another factor.

A very important role is played by design and manufacture in producing quiet bearings, but it must be emphasized that about 50 percent of bearing noise problems are caused by incorrect installation. This is particularly true in the case of roller-contact bearings. Alignment is extremely important.

5.1.9 Couplings. Couplings produce noise due to windage. This can usually be reduced effectively by installing a sheet metal enclosure or guard, lined with fiberglass or some other sound-absorbing material. Misalignment often produces noise and vibration in addition to the windage noise.

5.1.10 Jets. Jet noise increases as the eighth power of the discharge velocity. Therefore, substantial reductions can be obtained by reducing jet velocities.

5.1.11 Electric Motors. Electric motor noise is complex and is a combination of the following components:

1. Windage noise generated by the cooling air and resulting from turbulence. The turbulence is caused by the fan, and by the airstream, moving past stationary parts.

2. Windage noise generated by fan blades rotating close to a number of stationary members and producing siren or whistling noise. Fan noise can be reduced by using backward-curved blades. This is particularly important on 3600-rpm motors since fan noise varies approximately as the fifth power of the blade tip velocity. When acoustic silencers are used in the cooling air passages, noise reduction is greater if the number of fan blades is as large as possible, since high-frequency noise is easier to reduce than low-frequency noise.

3. Rotor-slot noise caused by rotating open slots. This component has a frequency equal to the number of slots times revolutions per second. It can be reduced by filling the slots with epoxy or some other material.

4. The most important source of noise due to electrical design is the combination of rotor bars and stator slots. This combination should be selected to provide minimum noise due to rotor and stator slot magnetomotive force interaction. Usually the combination is selected so that the number of exciting force-pole pairs is as high as possible, since the motor frame is stiffer in these modes.

5. Noise due to high flux density. This component, at twice line frequency, can be reduced by keeping the flux density as low as possible. Higher harmonics can be minimized by keeping air gaps as large as possible.

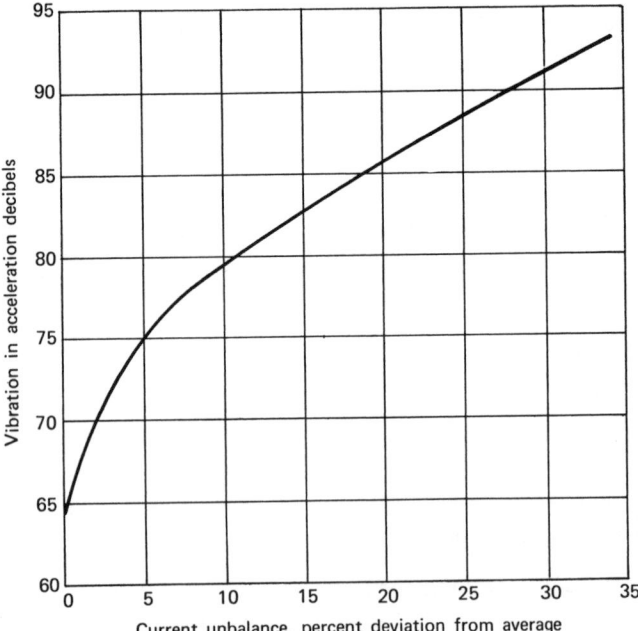

Figure 5.5

6. Mechanical factors responsible for excessive noise include misalignment of frame and end bells, which distort the frame and bearings, dynamic unbalance, noisy bearings, and structural resonances.

7. Unbalanced line currents in a three-phase power supply can increase the noise produced by a motor. Figure 5.5 shows the effect of line-current unbalance on structure-borne noise in terms of acceleration decibels, adB.

5.1.12 Center of Gravity Mounting. When it is possible machinery should be mounted so that the supports are symmetrical with respect to the center of gravity.

A machine can be considered to have six degrees of freedom—three translations and three rotations. When its supports are unsymmetrical with respect to the center of gravity, certain translation modes couple to rotational modes. Excessive vibration can excite resonances in various machine parts, panels, and foundation and can be radiated as noise. Mounting through the center of gravity reduces the probability of such excitation.

5.1.13 Improper Installation. Improper mounting of machinery can be the cause of fairly high sound levels. Heavy equipment should be mounted on adequate foundations or massive, rigid parts of a structure. It should never be mounted on lightweight beams or long unsupported spans.

It is better to place machines on separate foundations rather than on the same foundation, and it is desirable to keep machinery foundations separate from building floors or walls. Vibration beats are often caused by machines on the same foundation running at nearly the same speed. Beat frequency, in cycles per second or hertz, is equal to the difference in speed between the two machines, in revolutions per second.

5.1.14 Areas and Dimensions which can Radiate Certain Frequencies. Sound generated by a vibrating surface depends on the velocity of the surface motion and the area of the radiating surface. In general, an area with one dimension greater than one-quarter wavelength can effectively radiate sound at the frequency corresponding to that wavelength in air. Therefore, low-frequency sound radiation is limited to large surfaces, whereas any surface of more than a few square inches can radiate sound above 1000 Hz. Surfaces radiating certain low frequencies can often be divided into smaller areas, thereby reducing radiation efficiency.

Noise radiated from large areas can be reduced by providing openings or air leaks so that air can flow through and reduce pressure buildup.

5.2   Piping Systems

5.2.1   Effect of Flow Velocity in Pipes. Piping noise radiation in many instances exceeds noise radiated from machinery. In almost every case involving large compressors, pumps, and similar equipment, piping must be treated along with the machine in order to obtain the expected noise reduction. Remember, if piping noise equals machine noise, complete removal of the machine noise would reduce the original level by only 3 dB.

Flow noise in pipes increases with the velocity. Increasing the pipe diameter, therefore, reduces the noise. However, radiated noise also increases with the area of radiation, so this step increases noise. In general, the decrease due to reducing velocity is greater than the increase due to increasing pipe size.

Pipe noise can be transmitted from a pump or compressor into the piping system either mechanically or through the fluid, or both. If noise is transmitted mechanically, a vibration break in the piping can be very effective. When noise is transmitted through the fluid, the vibration break does not help. Either vibration isolation of the piping or piping lagging must be used.

Flow in piping systems can be laminar or turbulent. In laminar flow, individual particles move along parallel paths in the direction of motion. In turbulent flow, there is irregular, random motion.

Factors that determine whether flow is laminar or turbulent are pipe diameter D (in feet), fluid density p (in pounds per cubic foot), absolute viscosity μ (in pounds mass per foot-second), and flow velocity v (in feet per second). The variables are related by a dimensionless quantity, R, called Reynolds number.

$$R = \frac{Dvp}{\mu} \tag{5.7}$$

When the Reynolds number is below 1200, the flow is laminar. When it is more than 2200, the flow is turbulent. Between these two values, the flow can be either laminar or turbulent; 2000 is considered the dividing point.

In practically every industrial machinery piping system, the flow can be considered to be turbulent. This is probably the most important source of noise in centrifugal-compressor and pump piping systems.

5.2.2   Change in Pipe Diameter. An abrupt change in pipe diameter, such as where a pipe enters an expansion chamber, results in an acoustic impedance mismatch between the inlet pipe and the expansion chamber. Broad band sound in the inlet pipe is, therefore, partially reflected back toward the sound source. Therefore, the expansion chamber acts as a simple muffler.

High-velocity air flowing through an abrupt change in pipe diameter, from small to large, or vice versa, produces turbulence, however, and the turbulence noise can be transmitted directly through the pipe wall. To avoid this there should be a smooth transition from one pipe diameter to another.

5.2.3 Bends and Restrictions.

Sharp bends and restrictions in pipes create additional turbulence and more noise. The radius of bends should be at least five times the pipe diameter.

Turning vanes are of great importance in piping systems. Their purpose is to maintain correct flow conditions in going around bends. It must be emphasized that they are often a source of self-generated noise. They can introduce additional turbulence in the system, and they can vibrate at their own resonant frequencies and radiate them as pure tones.

When turning vanes are installed, the chord length and the separation between the vanes should not be uniform. They should be heavy enough to resist vibration, and they should not be identical in construction.

5.2.4 Von Karman Vortices.

When air or any other fluid flows by a cylindrically shaped obstacle, vortices are shed from the downstream side. The vortices alternate in direction, clockwise and counterclockwise, and this produces an alternating transverse force on the obstacle. The frequency of the alternating force, the diameter of the cylinder, and the velocity of flow are related by the equation

$$\frac{fD}{v} = 0.22 \tag{5.8}$$

where 0.22 is known as the Strouhal number.

Vibrations produced by the alternating force can be transmitted to the supporting structure and radiated as noise. Coincidence between the exciting force and one of the resonant frequencies of the structure can amplify it greatly.

EXAMPLE 5.2 Calculate the frequency of vibration to be expected if an obstacle 1 in. in diameter is placed in a pipe where the air flow is 100 ft/sec.

$$\frac{fD}{v} = 0.22$$

$$f = \frac{0.22v}{D}$$

$$f = \frac{0.22 \times 100}{0.0833} \doteq 264 \text{ Hz}$$

5.3   Centrifugal Machinery

5.3.1   Turbulent Flow inside Machinery.   Turbulence is the most important source of noise in centrifugal compressors. It is really a combination of two effects: vortex shedding and upstream turbulence. The boundary layer over each impeller blade is turbulent by the time it reaches the trailing edge. The

turbulent layers on the top and bottom surfaces produce a fluctuation in lift, and this fluctuation has a broad frequency spectrum. The application of a fluctuating force to a gas generates sound at the same frequency. Therefore, broad band noise is radiated. If the flow is turbulent when it enters a blade row, the turbulence is increased and the noise is greater. The noise inside the compressor, in the gas itself, can be quite high, and it is radiated directly through the compressor casing. It is almost impossible to eliminate turbulence noise by design.

Noise from this source also travels through the inlet and discharge piping through the gas. Vibration breaks in the piping do little to reduce it, and it is attenuated extremely slowly with distance. Noise reduction is accomplished by treatment after installation.

Von Karman vortices shed from the trailing edge of impeller blades also generate wide band noise inside the compressor, as shown in Fig. 5.6. In the case of axial-flow compressors, noise from this source increases with blade thickness. This is not necessarily true with centrifugal compressors. Noise

VON KARMAN VORTICES

Figure 5.6

from Von Karman vortex shedding is much higher in axial flow compressors if the air flow is such that the vortices leaving one blade are struck by the following blades.

In both centrifugal and axial compressors, noise due to turbulence is produced by any obstruction in the inlet or discharge that interferes with the air flow. This statement is also true for pumps. In fact, improper inlet conditions is one of the most important factors in centrifugal pump noise generation, and in the production of pressure pulsations in the fluid.

Bends inside pump and compressor casings, or changes in flow direction, are additional sources of turbulence noise. That is, for minimum noise generation, the entire internal flow path should be aerodynamically designed. This includes the following:

1. Keeping the flow path as large as possible, to decrease velocities, and as short as possible to reduce friction.
2. Removing all unnecessary obstructions.
3. Maintaining smooth surfaces, but not necessarily hand-finished.
4. Providing gradual changes in flow path cross section.

5.3.2 Inlet Conditions. Inlet and discharge conditions of centrifugal pumps and compressors have an extremely important part in the noise-generating ability of the machines, and in the pressure pulsations produced by them. The importance of bringing the fluid to the machine through a properly designed inlet piping system cannot be overemphasized, and the actual machine inlet is even more important.

It is sometimes thought that increasing the inlet and discharge areas reduces noise because velocities are less. On the contrary, if the suction opening is increased, it can actually create more noise rather than reduce it. In order to obtain proper entrance, the fluid should be as near as possible to the impeller center line. This, of course, indicates a small inlet. When the fluid enters near the center line on its way to the impeller vane, there is relatively low shock and turbulence. If the suction opening is enlarged, the fluid is admitted farther up on the impeller vane where the linear speed is higher. The sudden change from low velocity in the large inlet to high velocity part way up on the vane causes shock, turbulence, and increased noise.

Reducing the flow velocity in the discharge piping is an effective way to reduce piping noise. However, if the discharge opening is increased by simply relocating it with respect to the cutwater, then it has not really changed anything. It is usually more satisfactory to accomplish the same thing by using a properly designed pipe increaser to connect the pump or compressor discharge to the discharge piping.

5.3.3 Impeller Rotation Speed.   Speed of rotation has a definite effect on noise. For any particular design the sound level increases anywhere from 20 to 50 times the logarithm of the speed ratio. At lower speeds, up to about 7500 RPM, centrifugal compressor noise increases by an amount equal to 20 times the logarithm of the speed ratio. At speeds above this, the increase approaches 50 times the logarithm of the speed ratio.

The increase in sound with speed applies to the overall noise and the component of highest level, usually at blade-passing frequency or blade-rate frequency. The increase at other frequencies is not as great, and it can be of the order of 10 to 15 times the logarithm of the speed ratio.

Centrifugal pump noise increases in a similar fashion except not quite as rapidly at high speed. There the increase approaches 40 times the logarithm of the speed ratio.

These same relations apply to impeller tip speeds, as well as to rotational speed, but, in general, less noise is produced by large-diameter, slow-speed machines than by small-diameter, high-speed units, even though the impeller tip speeds are the same in both cases. There are several reasons for this:

1. Not all the turbulence is produced by the impeller. Some is generated by flow through internal passages. The larger areas in slow-speed machines offer less restriction and less turbulence.

    Although internal passages should be smooth, extra care in producing fine interior finish in centrifugal machines does not appear to be justified. There is no detectable difference in sound level when passages are hand-finished. Perhaps if all other sources of noise were reduced to low enough levels, an additional improvement could be obtained by hand-finishing.

2. Mechanical forces due to unbalance are proportional to the square of the speed. Low speed produces less structure-borne sound and less excitation of structural resonances.

Figure 5.7 shows how noise level varies with speed for one class of centrifugal compressor.

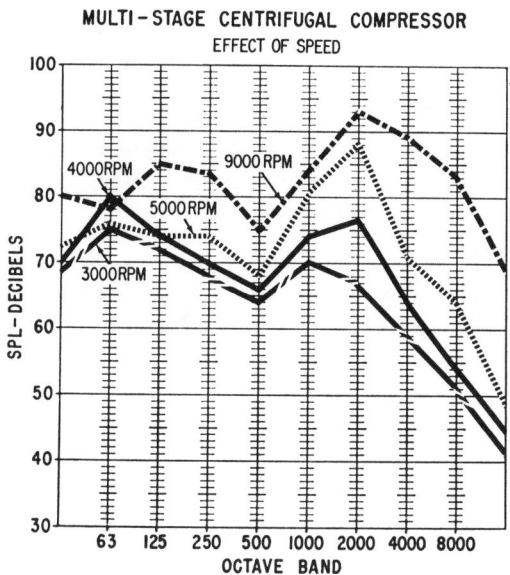

Figure 5.7

5.3.4 Blade-Passing Frequency.
Every time an impeller blade passes a given point, the air or fluid at that point receives an impulse. Therefore, that point receives impulses at a frequency equal to the number of blades times revolutions per second.

In axial-flow compressors this blade-passing frequency is the major component in the generated sound spectrum. In centrifugal compressors and centrifugal pumps it is usually strong, but not as pronounced as it is in the case of axial compressors.

5.3.5 Blade-Rate Frequency.
Blade-rate frequency is often the major noise component in diffuser-type machines. This terminology is used to differentiate it from blade-passing frequency, which is equal to blades times revolutions per second.

Blade-rate frequency is calculated by the equation

$$f = \frac{N_R \times N_S}{K} \times \text{rps} \tag{5.9}$$

where f = the frequency, in Hz,

N_R = the number of rotating (impeller) blades,

N_S = the number of stationary (diffuser) vanes, and

K = the highest common factor of N_R and N_S.

As the impeller rotates, certain combinations of rotating and stationary vanes coincide during each revolution. The number of times that rotating and stationary vanes coincide is equal to the product of the number of rotating and stationary vanes divided by the highest common factor. Each time there is coincidence, the number of rotating blades matching stationary vanes is equal to the greatest common factor.

This is shown in the following example:

EXAMPLE 5.3   Calculate the blade-rate frequency if the number of rotating blades is 6, the number of stationary vanes is 9, and the speed is 6000 RPM.

$$f = \frac{N_R \times N_S}{K} \times \text{rps}$$

$$f = \frac{6 \times 9}{3} \times \frac{6000}{60} = 1800 \text{ Hz}$$

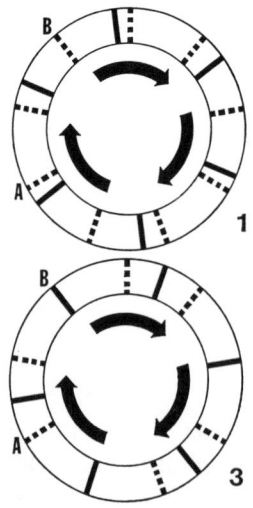
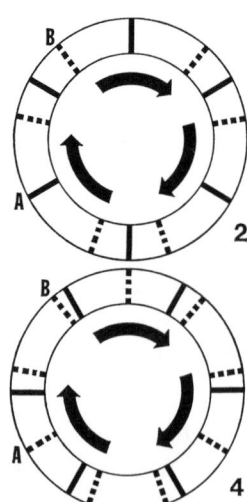

Figure 5.8

There are eighteen times in each revolution when rotating blades line up with stationary vanes, and each time this happens, three rotating blades match three stationary vanes. Therefore, the frequency is eighteen times the revolutions per second, and the pulses are three times as strong as they would be with a single blade coincidence.

This is shown in Fig. 5.8. Rotating blades are solid lines and stationary vanes are dotted. As the impeller rotates, the blades come into coincidence with stationary vanes. The first coincidence is reached at the three points A. This is followed by coincidence at three points B, and so on.

It can be seen that it is better to use unequal numbers of rotating and stationary vanes, and that prime numbers are best because they have no common factor except 1. This keeps the frequency high and the pulse strength low. It is easier to control high-frequency noise than low-frequency noise.

5.3.6 Impeller-Diffuser Distance. Increasing the radial distance between impeller blades and diffusers reduces noise. It is particularly effective in reducing the blade-passing frequency and blade-rate components. Unfortunately this procedure also decreases efficiency, but for close initial spacing, the noise decreases more rapidly than the efficiency. That is, noise increases rapidly as the spacing becomes smaller and smaller.

5.3.7 Uneven Blade Spacing. Various theories have been offered on the advantages of uneven blade spacing. Supposedly this would suppress the generation of blade-passing and blade-rate frequencies. Unfortunately, they cannot be spaced with enough inequality to be really effective, and other problems are encountered, such as upsetting the dynamic balance and increasing manufacturing costs.

5.3.8 Blade Loading. The relation between generated noise in centrifugal machinery and the number of impeller blades is not definite. If the number of blades is very small, and they are narrow, then increasing the number decreases noise, and the number of audible higher harmonics is reduced. If the number of blades is doubled, the overall noise should decrease about 3 dB, that is, by $10 \log N_2/N_1$. On the other hand, there is evidence that high-solidity blade rows generate more noise because of their greater surface area.

In general, increasing the number of stages in a centrifugal machine decreases generated noise because the work per stage is reduced. In some compressors the sound is reduced about $25 \log N_2/N_1$, where N_2/N_1 is the ratio of the number of stages.

5.3.9 Impeller Vibration Induced by Steady Force. It is well known that a rotating impeller or disk, such as a centrifugal compressor impeller or a steam turbine wheel, can be made to vibrate by an alternating force corresponding to one of its resonant frequencies.

Not as well known, however, is the fact that a dangerous type of vibration can be induced by the application of a small steady force, such as that from some irregularity in one of the nozzles of a steam turbine.

This can occur when the wheel is rotating at one of its "disk critical speeds," that is, when the backward-traveling wave speed for that particular frequency is identical to the forward speed of the wheel. When this happens the vibration wave in the wheel stands still in space.

Usually the vibration is balanced with respect to the wheel, and, therefore, it cannot be detected on the shaft, even though large amplitudes of vibration may be present in the wheel.

Sound at the same frequency as the vibration can be radiated, but noise is not the most important consideration in these cases, and noise control is not the solution. Vibration at disk critical speeds can cause impeller failure in a very short time, and although a discrete tone may be produced, there may not be enough time to isolate it or even identify it. The solution is to change the vibration frequency of the wheel, or move the operating speed away from the disk critical speed.

5.3.10 *Effect of Gas Molecular Weight.* The molecular weight of the gas in centrifugal compressor systems has a pronounced effect on the generated

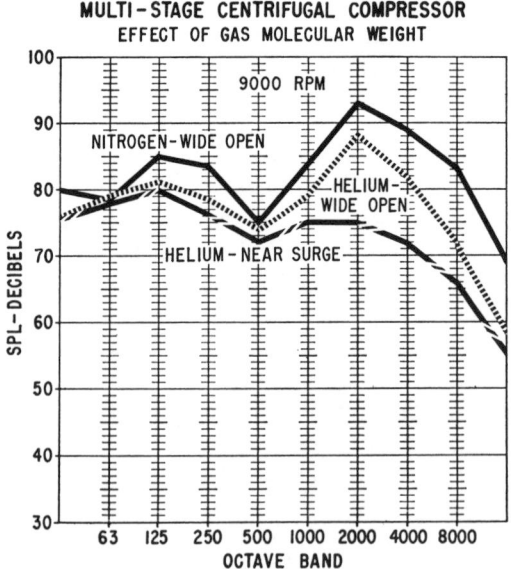

Figure 5.9

noise. Very little test data are available concerning this, and the mathematical relations are difficult to establish. However, there are enough data to show that more noise is produced with high molecular weight gas than with low molecular weight gas. Figure 5.9 shows the decrease in octave band sound pressure levels for one particular type of centrifugal compressor, running at 9000 RPM, when helium (molecular weight = 4) is used instead of nitrogen (molecular weight = 28).

5.3.11 Effect of Operating Point. Figure 5.9 also shows how centrifugal compressor noise varies with the operating point. In general, less noise is produced when a centrifugal compressor or centrifugal pump is operated at its rated conditions. But, when the flow is reduced from wide open to a point farther back on the operating curve, and near to the point where surge begins, the noise decreases. When the surge point is reached, however, noise increases to a high level, and it often sounds as though a gun is being fired. Figure 5.10 shows another example of this, taken at 4000 RPM, with nitrogen in the system. No noise data are shown during actual surge since a compressor should not be held at that point while octave band sound pressure levels are being measured.

5.3.12 Mass Flow. Mass flow and discharge pressure both have a profound effect on the noise produced by a compressor. This is also shown on Figs. 5.9 and 5.10. Even though the machine is being operated well out on its

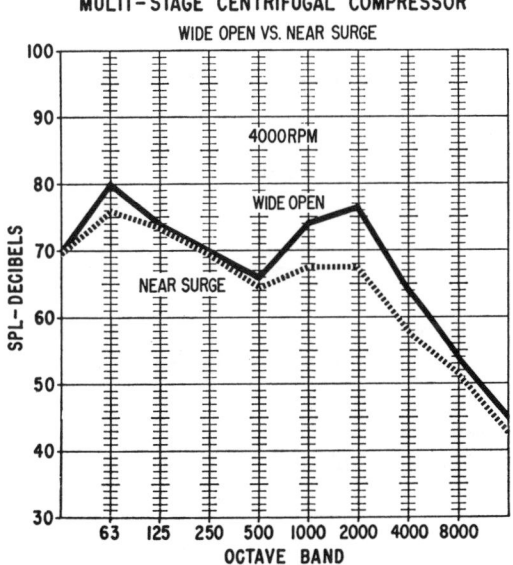

Figure 5.10

head-capacity curve, and not near its surge point, there is a decrease in noise as the mass flow is reduced. Of course, when the surge point has been reached, noise increases rapidly.

The effects of mass flow and discharge pressure are difficult to predict analytically, and they differ from one class of compressor to another. For any particular class, however, the effects of changes in flow and pressure can be calculated once certain controlled sound tests have been made to establish a base line.

5.3.13 Pressure Pulsations. Severe piping vibration has been observed at some centrifugal pump installations during certain operating conditions. The vibration is usually accompanied by an increase in noise.

There are a number of causes of piping vibration. For example, a section of pipe can be excited by one of the forcing frequencies in the pump, usually blade-passing frequency or blade-rate frequency. The resonant frequency of a pipe is a function of its length and the velocity of sound in the fluid. Open-ended and closed-ended pipes respond to a fundamental frequency and all its higher harmonics.

The wavelength of the fundamental wave is equal to twice the length of the pipe. Therefore, the frequency of the fundamental and harmonics is given by the following equation:

$$f = \frac{V}{2L}, \quad \frac{2V}{2L}, \quad \frac{3V}{2L}, \quad \frac{4V}{2L}, \quad \text{and so on} \qquad (5.10)$$

where L = length of pipe, in feet,
$\quad V$ = velocity of sound in the fluid, in feet per second, and
$\quad f$ = frequency, in hertz.

The wavelength of sound in a pipe open at one end and closed at the other is four times the length of the pipe. In this case, excitation can be produced by the fundamental frequency and all the odd harmonics. That is,

$$f = \frac{V}{4L}, \quad \frac{3V}{4L}, \quad \frac{5V}{4L}, \quad \frac{7V}{4L}, \quad \text{and so on} \qquad (5.11)$$

Sometimes in practice it is not easy to determine whether a particular section of pipe acts as one with both ends open or as one open at one end and closed at the other. Changes in pipe diameter, bends, direction, entrance into various system components, and valves cause partial reflection of sound waves, and they have an effect on resonant frequencies.

Other factors that can cause pipe vibration are rotating stall, before surge actually begins, and cavitation.

The principal cause of excessive vibration in large pumping systems, such as boiler feed pumps, is operation of the pump on an unstable part of the head-capacity curve. The vibration usually occurs at a low frequency, although higher frequency components almost always are present also.

The conditions necessary for this to happen are the following:

1. The mass of water must be able to move back and forth. This condition exists in boiler feed pump systems where both ends of the system can be considered as free surfaces.
2. There must be a spring, or elasticity, in the system that can store and release energy. This condition prevails in the elasticity of the piping system itself, the steam in the boiler, and to a lesser extent, the compressibility of the water in the system.
3. There must be an exciting force at some particular frequency to induce the surges. This oscillating force also can exist.

If the pump is operating on a stable part of the head-capacity curve, such as point A in Fig. 5.11, any tendency to reduce flow is accompanied by an

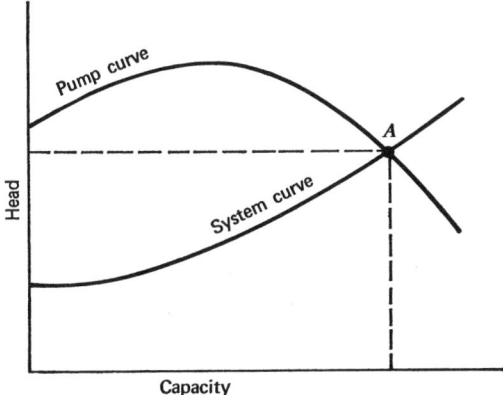

Figure 5.11

increase in pump head, and operation at point A is maintained. A tendency to increase flow results in a decrease in head, and stable operation at point A is again held.

If the pump is operating on an unstable part of the head-capacity curve, and the flow is absolutely steady, no head-capacity swing should be produced, and no vibration or noise should result from this cause. However, the flow is never absolutely steady, and a momentary shift in the flow, one way or the other, can initiate the surge. For example, if the head-capacity curve has a decided droop, or positive slope, at low flows, as shown on Fig. 5.12, and the pump is operating at point B, a sudden decrease in flow causes the head produced by the pump to be below the line pressure, and the flow through

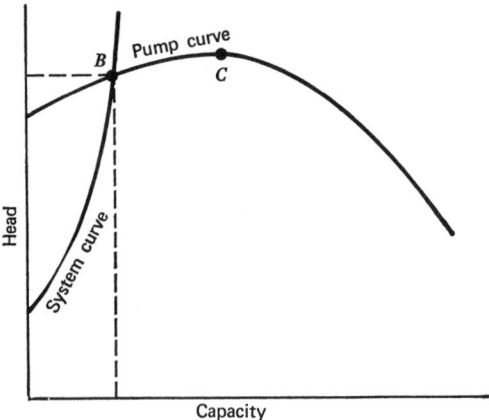

Figure 5.12

the pump is decreased further. That is, a decrease in flow results in less head and still lower flow. The flow continues to decrease until the system pressure drops below the pump pressure, at which time the pump discharges into the system again. The head then continues to increase with the flow until point C is reached. But, if the system requires only the capacity at point B, the flow again starts to decrease and the cycle repeats.

If the frequency of the surges coincides with one of the frequencies of the piping system, severe vibration and noise can be produced.

5.4   Reciprocating Machinery

Noise produced by reciprocating machinery is usually a multiple of the piston movement. The noise is generated by both aerodynamic, or fluid noise, and mechanical sources.

5.4.1   Inlet and Exhaust Noise.   Both the inlet and exhaust are major noise sources in reciprocating machinery such as compressors and engines. The pulsating flow can be silenced effectively by a chamber-type silencer or snubber. Inlet noise is usually the second highest noise source, and it is exceeded only by that of the exhaust. Engine exhaust usually occurs at the fundamental engine firing frequency. It is usually 8- to 10-dB higher than the inlet noise. Both inlet and exhaust noise increase with load, and they are affected considerably by the size of valves, timing, and the way the ports are constructed. Inlet noise is also affected by the construction of the exhaust system.

5.4.2 Combustion Noise. Combustion noise is predominant after the inlet and exhaust have been silenced. It is dependent on the pressure rise in the cylinders. The theory of how noise is produced by the combustion process is not very well understood. The actual pressure rise does not seem to be as important as the rapidity of the rise, and it is known that combustion noise can be varied by as much as 10 dB by changing the form of the cylinder pressure rise. This is an appreciable change in loudness, and it indicates that investigations along these lines could prove worthwhile.

The cylinder pressure wave form contains quite a few higher harmonics in addition to the fundamental frequency. The low-frequency components of the cylinder pressure wave are determined chiefly by the peak pressure. The higher harmonics are affected more by the shape of the pressure diagram.

Pressure curves of diesel engines differ from those of gasoline engines, particularly from 800 to about 3000 Hz. The difference is due mainly to the ignition process. In the gasoline engine, combustion is initiated by a spark, and the flame propagates comparatively slowly. In diesel engines, the ignition is rapid, causing a rapid rise in pressure and a rather broad peak from about 800 to 2000 Hz.

The spectrum of gasoline engines is different. The components in the 800- to 2000-Hz range are usually lower in amplitude and the greatest peaks are in the 400 to 600-Hz range. While the sound levels change with load, the shape of the spectrum, and the components present in the total noise, show a marked difference between diesel and gasoline engines. Diesel engine spectra change very little from no load to full load. There is a considerable change in the shape of gasoline engine sound spectra.

5.4.3 Mechanical Noise. High-level mechanical noise, in many instances, masks the combustion noise. These mechanical noise sources include the following:

1. Crank-connecting rod.
2. Fuel injection system.
3. Valves.
4. Excessive clearance in cross heads.
5. Inadequate lubrication.
6. Misalignment.
7. Bent parts.
8. Excessively tight packing.
9. Loose parts.
10. Improper mounting.

In general, the combined effect of all the sources increases engine noise in proportion to about 10 times the logarithm of the horsepower. That is, the increase is about 3 dB per doubling of the horsepower. The noise increases about 30 times the logarithm of the speed ratio, that is, about 9 dB per doubling of speed.

Inertia forces are major exciting forces causing vibration and noise in reciprocating machinery. These forces are due to the motion of the pistons and related parts and the imbalance of the connecting rod and crank mechanism. The forces produced by unbalanced masses also appear in the rotating parts of the machine as both static and dynamic unbalance.

Impacts in the crank-connecting rod system, particularly the knocking of the pistons against the cylinder liners during crossover, have been found to be an important internal noise source in reciprocating engines. This source of noise is also a major one in reciprocating compressors and pumps.

During each revolution of the crankshaft, the piston shifts several times from one side to the other, moving in the plane of the connecting-rod motion. The gap between the piston and the cylinder liner permits the piston to move with a certain velocity in the transverse direction, impacting against the wall of the cylinder. These knocks produce an intense vibration of the cylinder walls at their resonant frequency, resulting in a noise level 3- to 9-dB higher than any other internal noise.

Factors that determine the noise level of the impacts are the following:

1. Speed.
2. The amount of clearance in the bearings.
3. Weight of the piston and connecting rod.
4. The force acting on the piston.
5. The ratio of the crank radius to the length of the connecting rod.
6. The material in the cylinder block and cylinder head.
7. The thickness of the cylinder block.
8. The number of cylinders.
9. The viscosity of the lubricant.

The mechanical noise produced by piston impacts can, therefore, be reduced by reducing clearances, piston and connecting rod weights, and speed.

Torsional vibration can cause problems. Naturally, machinery is designed to operate at speeds removed from any lateral or torsional critical speeds. Certain engines, however, can operate fairly close to a torsional critical speed. This can produce additional bearing impacts and noise, usually of a low frequency.

The fuel system of engines is often an additional source of noise. The injection process, impacts of valves on their seats, and driving gear tooth contact noise all combine in this source. Usually this noise is considerably less than piston impact noise.

5.5   Rock Drill Noise

Rock drill and paving breaker noise contain components from the air exhaust, vibration of the drill steel or moil point, noise radiated from the drill body, internal noise from valves, ratchets, rotation and feed mechanism, chuck rattle, and noise from penetration of the rock.

5.5.1 Exhaust Noise.   The air exhaust is the largest component in rock drill noise, and usually it is 5- to 10-dB higher than the next highest component, that produced by vibration of the drill steel. Exhaust noise is fairly broad in nature, with high levels in all octave bands. It is futile to try to reduce rock drill noise without first reducing the air exhaust component.

Exhaust noise can be reduced by chamber-type or absorptive-type mufflers that provide reductions of about 5 to 8 dB. Another effective way to reduce the exhaust component is to pipe it away with a long hose, but this is often unacceptable to users since it means handling another hose.

5.5.2 Steel Vibration.   Next in importance is the noise produced by vibration of the rock drill steel. When the piston strikes the end of the drill steel the complex vibration causes sound to be radiated at many different frequencies. If the piston strikes the steel perfectly axially, the vibration would be in the longitudinal direction only, at a fundamental frequency and higher harmonics, which are related to the length of the steel and the velocity of sound in the material. Because the striking surfaces are never perfectly square, and because there is always some clearance between piston and cylinder, and between the steel and the front head, bending occurs. Bending mode frequencies are dependent on the length and cross section of the steel and the modulus of elasticity. Noise produced by bending vibration is at lower frequencies than that caused by longitudinal vibration.

Tests show that steel-radiated noise must be reduced along with air exhaust noise if any significant reduction in overall rock drill noise is to be accomplished. Figure 5.13 illustrates this. Curve *A* shows air exhaust noise only measured on a large, powerful rock drill. The exhaust was conducted 100 ft away from the drill by means of a hose. These octave band levels are equivalent to 117 dB*A*.

Curve *B* shows the octave band sound pressure levels measured at the same location when a good muffler was installed on the end of the hose. The muffler produced an insertion loss of 21 dB, leaving only 96 dB*A*. This would indicate that a substantial noise reduction, 21 dB*A*, could be obtained by applying a high-quality muffler to the exhaust.

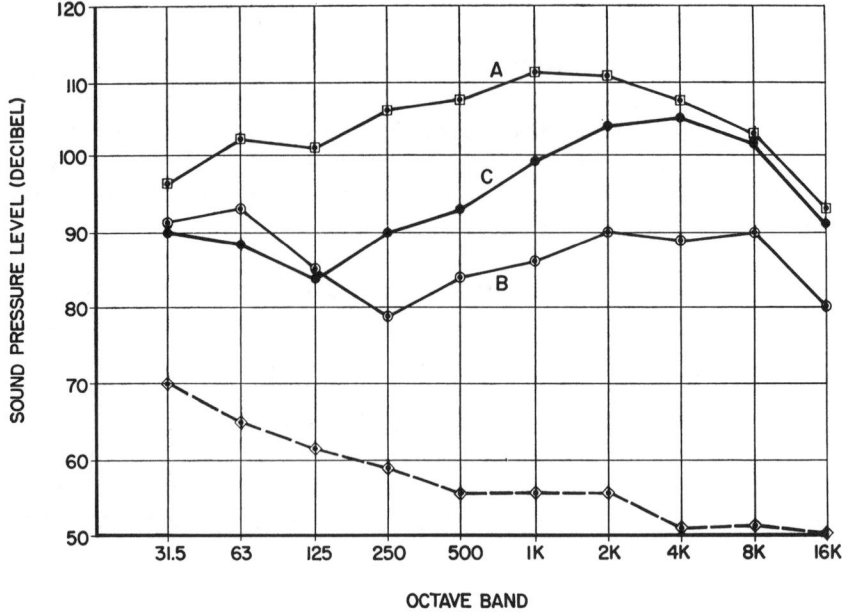

A ▣ AIR HOSE - ISOLATED FROM DRILL- WITHOUT MUFFLER
B ◉ AIR HOSE - ISOLATED FROM DRILL- WITH MUFFLER
C ● STEEL NOISE AND CASING RADIATION
◈ BACKGROUND NOISE

Figure 5.13

However, when octave band levels are measured back at the drill itself, with the exhaust noise piped away and muffled, the sound level is found to be 110 dB*A*, a 7-dB*A* reduction compared to the original noise, as shown on curve *C*. This shows why all rock drill mufflers produce noise reductions of only 5 to 8 dB*A*.

When damping is applied to the drill steel, additional reductions of 2 or 3 dB can be obtained. Further improvements can be effected by more sophisticated damping treatments, but this can add substantially to the cost of the steel.

5.5.3 Mechanical Noise.   Noise radiated directly by the drill body, including internal impact noise, and components from valves, rotation and feed mechanisms, chuck rattle, and other lesser contributors, can be reduced by enclosures or shields to provide the necessary transmission loss. This technique is easier to use on large drills than on small hand-held ones since it can add quite a bit to the size and weight.

CHAPTER 6

MACHINERY SOUND CONTROL

The best sound-control technique is to prevent the generation of noise, or at least to reduce the level of generated noise. Although this is possible in many cases, there are times when it is better and more economical to reduce noise after a machine has been built.

When machinery has been designed and manufactured for maximum performance, internal changes to reduce noise can also reduce efficiency. One or two points in efficiency is often too great a price to pay for noise reduction, and customers are unwilling to accept it. In such cases it is better to leave the machine alone and reduce its radiated noise after it has been installed. For example, a decrease of 1 or 2 percent in the efficiency of a 70,000-hp boiler feed pump represents a sizeable increase in the annual power cost to operate it. This would easily pay for external, supplementary noise control, applied after the machine has been installed.

When this is necessary, the first step is to find where the objectionable noise is coming from and the path it takes to the ear of the listener. The next step is to decide on how to reduce the sound to the design level.

Noise can be radiated directly from a vibrating machine part to the ear of the listener, or it may cause vibration in some other part, which in turn sends out additional sound waves. These may actually be amplified by structural resonances excited by either airborne sound or by structure-borne vibration.

A knowledge of certain fundamental acoustic principles is necessary in order to apply effective noise reduction measures.

In general, there are five basic methods used to reduce noise: sound absorption, sound isolation, vibration isolation, vibration damping, and muffling. These five methods do not overlap. The most effective sound control, at minimum expense, can be accomplished if they are understood.

6.1 Sound Absorption

Earlier, in the discussion of semireverberant fields, sound absorption coefficient was defined as the ratio of the sound energy absorbed by a surface to the energy incident on the surface.

Materials that have high absorption coefficients usually have soft, porous surfaces. When sound waves strike these surfaces, air flows in and out of the minute pores in the material because of the pressure changes produced by the sound. Frictional forces convert the sound energy into heat, although the actual amount of energy is small.

It was shown that reverberant sound in a room containing a noise source can be reduced by increasing the absorption in the room. This can be done by increasing the absorption coefficient of the material, or by increasing the area on which absorbing material is placed.

Energy is taken out of the sound waves each time they pass through the material, as they are reflected many times from wall to wall.

6.2 Sound Isolation

A common method for reducing machinery noise is to place a barrier or wall between the noisy machine and the listener. The effectiveness of such a barrier is described by its transmission coefficient.

Sound transmission coefficient of a partition is defined as the fraction of incident sound transmitted through it.

Sound transmission loss is a measure of sound-isolating ability, and it is equal to the number of decibels by which sound energy is reduced in transmission through a partition. By definition, it is 10 times the logarithm to the base 10 of the reciprocal of the sound transmission coefficient. That is,

$$TL = 10 \log \frac{1}{\tau} \tag{6.1}$$

where TL = the transmission loss, in decibels, and
τ = the transmission coefficient.

EXAMPLE 6.1   The sound transmission coefficient of a wall is 0.001. Calculate the sound transmission loss.

$$TL = 10 \log \frac{1}{0.001}$$

$$= 10 \log 1000$$

$$= 10(3.0)..$$

$$= 30 \text{ dB}$$

Transmission of sound through a rigid partition or solid wall is accomplished mainly by the forced vibration of the wall. That is, the partition is forced to vibrate by the pressure variations in the sound waves.

Under certain conditions porous materials can be used to isolate high-frequency sound, and in general, the loss provided by a uniform porous material is directly proportional to the thickness of the material. For most applications, however, sound-absorbing materials are very ineffective sound isolators; they have the wrong characteristics. They are porous, instead of airtight, and are lightweight, instead of heavy.

In the case of nonporous materials, the transmission loss is determined chiefly by its weight per square foot of surface area and how well it is sealed. That is, heavy partitions are better noise isolators than lightweight ones, provided they are not porous and all cracks and openings are sealed. Transmission loss is affected also by dynamic bending stiffness and internal damping.

6.2.1 Single Panel.   The simplest type of sound isolating barrier is a single, homogeneous, nonporous partition. In general, the transmission loss of a single wall of this type is proportional to the logarithm of the mass. Its isolating ability also increases with frequency, and the approximate relationship is given by the following equation:

$$TL = 20 \log W + 20 \log f - 33 \tag{6.2}$$

where TL = the transmission loss, in decibels,
      W = the surface weight, in pounds per square foot, and
      f = the frequency, in hertz.

This means that the transmission loss increases 6 dB each time the weight is doubled and 6 dB each time the frequency is doubled. In practice, both of these will be less than 6 dB.

6.2.2 Stiffness.   At low frequencies, transmission loss is controlled mainly by stiffness, and it increases about 6 dB per doubling of stiffness. Transmission loss in this range decreases 6 dB per octave as the frequency increases.

This indicates that at low frequencies, panel stiffness should be increased as much as possible. Mass and damping are unimportant in this region.

6.2.3 Resonance. At frequencies somewhat higher than those in the stiffness-controlled region, a series of plate-type resonances in the panel cause wide fluctuations in the transmission loss. Vibration of the panel results in noise radiation at these frequencies, and the transmission loss is limited chiefly by whatever damping is present in the panel.

6.2.4 Mass. Above the first series of resonances, transmission loss is controlled by the mass of the panel. Although equation (6.2) indicates that it increases 6 dB for each doubling of the surface weight, it turns out in practice that transmission loss increases about 4 to 5 dB each time the surface weight is doubled. This rate of increase continues over a range of frequencies from 2 or 3 times the lowest resonance frequency up to a certain "critical frequency."

6.2.5 Wave Coincidence. It would be expected that a high amplitude of vibration would be induced in a panel or wall if one or more of the frequency components in a sound wave matched one or more of the normal modes of vibration of the panel. It may be somewhat surprising however to find that for every frequency above a certain critical frequency, there is a particular angle of incidence at which the panel vibrates as though it were at resonance.

The reason for this response is that above the critical frequency, the wavelength of the bending wave in the panel can become equal to the wavelength of the sound wave in air, projected on the panel. When this happens a strong coupling between the panel and the airborne sound wave is obtained. This condition is called wave coincidence.

When a sound wave is incident on a panel at the frequency and angle where wave coincidence occurs, the panel vibrates just as though it were at resonance, and it becomes nearly transparent to sound.

Figure 6.1 shows how this is brought about. The wavelength of the bending wave in the panel is λB. If the wavelength of the sound wave in air is λ, and it impinges on the plate at an angle Φ_0 so that $\lambda/\sin \Phi_0$ is equal to λB, the intensity of the transmitted wave approaches the intensity of the incident wave. The frequency for which $\lambda = \lambda B$ is called the critical frequency.

Under this condition the panel vibrates at an amplitude nearly equal to the amplitude of the air motion in the incident wave. This causes the panel to radiate a transmitted wave with almost the same amplitude, and at the coincidence angle Φ_0. That is, the panel radiates a wave that is almost as intense as the exciting wave.

The bending wave in the panel is a traveling one, and it is not the same as

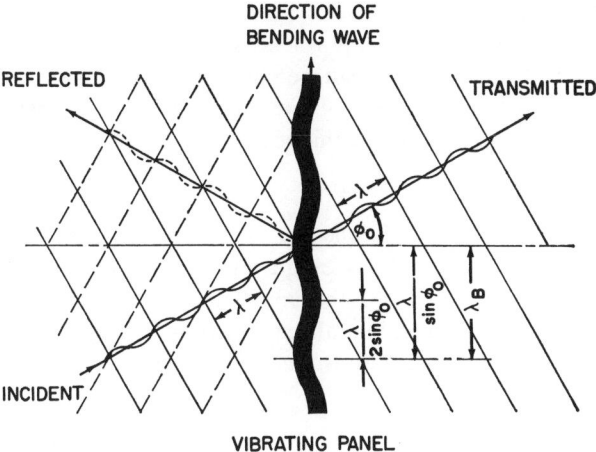

Figure 6.1

the usual type of panel vibration where the waves are stationary and fixed nodes can be located. It can be seen that wave coincidence can occur only when the wavelength of the sound in air is less than the wavelength of sound in the plate.

Figure 6.2 shows critical frequencies (f_c) for various materials plotted as a function of their thickness (h). Note that for steel or aluminum $\frac{1}{2}$-in. thick,

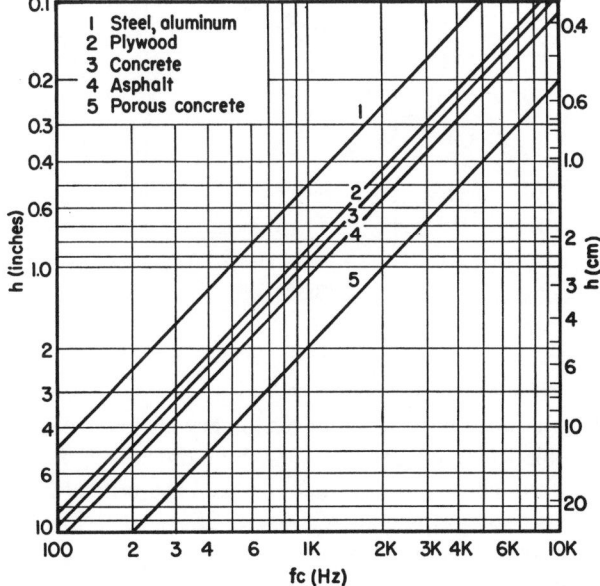

Figure 6.2

the critical frequency is 1000 Hz. Wave coincidence can occur at all frequencies above this. If the plate thickness is only 0.1 in., the critical frequency is raised to 5000 Hz. It can be shown that the critical frequency is proportional to the square root of the mass, in pounds per cubic foot, and inversely proportional to the thickness.

6.2.6 Region above Critical Frequency.   At frequencies above the critical frequency, panel stiffness and damping again become important. The transmission loss again increases with frequency at a rate somewhat higher than that in the mass-controlled region, about 8 to 10 dB/octave. Figure 6.3 shows how transmission loss varies with frequency in these various regions.

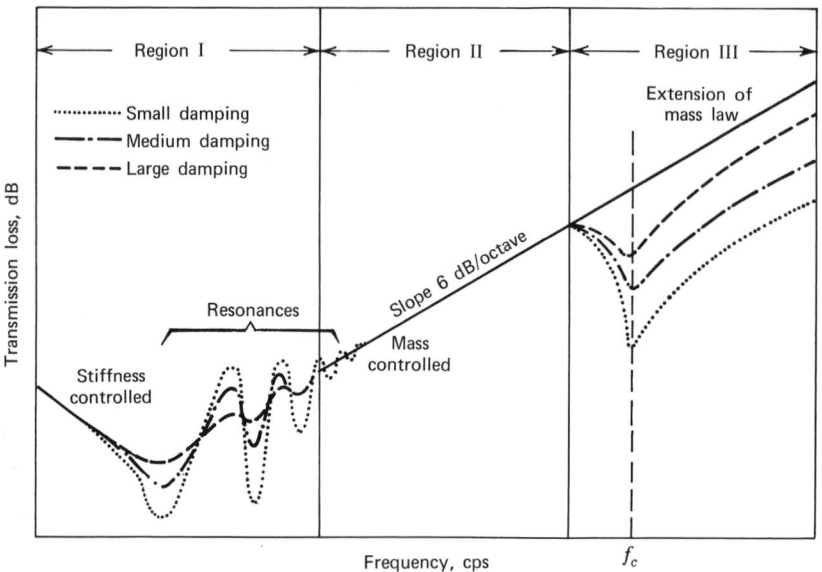

Figure 6.3

The paragraph above indicates that machinery enclosures should be made as stiff as possible for low frequencies, with as much weight as possible for medium-range frequencies, low stiffness to keep the critical frequency high, and high stiffness and damping at frequencies above the critical frequency. It is obvious once again, as in many other engineering problems, that a compromise must be made to obtain the optimum design.

Table 6.1 shows the transmission loss of some common building materials.

6.3 Composite Panel

Many walls or sound barriers are made of several different materials, each with a different transmission coefficient. For example, machinery enclosures are commonly constructed of sheet steel, but they may have a glass window to observe instruments inside the enclosure. The transmission coefficient of steel is, of course, different from the corresponding value of glass. Another example is when there are necessary cracks or openings in the enclosure where it fits around a rotating shaft. In this case, the transmission loss of the opening is zero, and the transmission coefficient is 1.0.

Naturally the effectiveness of such a wall is related to both the transmission coefficients of the materials in the wall and the areas of the sections. A large area transmits more noise than a small one made of the same material. Also it can be seen that more noise can be transmitted by a large area with a relatively small transmission coefficient than by a small one with a comparatively high transmission coefficient. On the other hand, a small area with a high

TABLE 6.1 TRANSMISSION LOSS OF BUILDING MATERIALS

Item	TL
Hollow-Core Door ($\frac{3}{16}$-in. Panels)	15
$1\frac{3}{4}$-in. Solid-Core Oak Door	20
$2\frac{1}{2}$-in. Heavy Wood Door	25–30
4-in. Cinder Block	20–25
4-in. Cinder Block, Plastered	40
4-in. Cinder Slab	40–45
4-in. Slab-Suspended Concrete, Plastered	50
Two 4-in. Cinder Blocks—4-in. Air Space	55
4-in. Brick	45
4-in. Brick, Plastered	47
8-in. Brick, Plastered	50
Two 8-in. Cinder Block—4-in. Air Space	57

transmission coefficient can ruin the effectiveness of an otherwise excellently designed enclosure. Another way of stating the problem is that both transmission coefficients and areas must be controlled carefully.

The average transmission coefficient of a composite panel is

$$\bar{\tau} = \frac{\tau_1 S_1 + \tau_2 S_2 + \tau_3 S_3 + \cdots \tau_n S_n}{S_1 + S_2 + S_3 + \cdots S_n} \tag{6.3}$$

where τ_1, τ_2, τ_3, and so on, = the various transmission coefficients of areas S_1, S_2, S_3, and so on.

The transmission loss of a composite panel can be calculated from this equation as follows:

Let the panel consist of two areas, S_1 and S_2, with corresponding transmission coefficients τ_1 and τ_2. Then the average transmission coefficient is

$$\bar{\tau} = \frac{\tau_1 S_1 + \tau_2 S_2}{S_1 + S_2}$$

The transmission loss of the composite panel is

$$TL_c = 10 \log \frac{1}{\bar{\tau}} = 10 \log \frac{S_1 + S_2}{\tau_1 S_1 + \tau_2 S_2}$$

Let $S_2/(S_1 + S_2) = K$, or $S_2 = KS_1 + KS_2$. Then

$$S_2 - KS_2 = KS_1$$

$$S_2(1 - K) = KS_1$$

$$S_2 = \frac{KS_1}{1 - K}$$

Therefore,

$$TL_c = 10 \log \left[\frac{S_1 + (KS_1/1 - K)}{\tau_1 S_1 + (\tau_2 KS_1/1 - K)} \right]$$

$$= 10 \log \left[\frac{1 + (K/1 - K)}{\tau_1 + (K\tau_2/1 - K)} \right]$$

From equation (6.1)

$$TL_1 = 10 \log \frac{1}{\tau_1}$$

and

$$TL_2 = 10 \log \frac{1}{\tau_2}$$

The difference in transmission loss between the two areas is

$$TL_1 - TL_2 = 10 \log \frac{1}{\tau_1} - 10 \log \frac{1}{\tau_2}$$

$$= 10 \log \frac{1/\tau_1}{1/\tau_2} = 10 \log \frac{\tau_2}{\tau_1}$$

The difference in transmission loss between area S_1 and the composite area is

$$TL_1 - TL_c = 10 \log \frac{1}{\tau_1} - 10 \log \left[\frac{1 + \dfrac{K}{1-K}}{\tau_1 + \dfrac{K\tau_2}{1-K}} \right]$$

$$= 10 \log \frac{1}{\tau_1} - 10 \log \left[\frac{\dfrac{1-K+K}{1-K}}{\dfrac{\tau_1 - K\tau_1 + K\tau_2}{1-K}} \right]$$

$$= 10 \log \frac{1}{\tau_1} - 10 \log \left[\frac{1}{\tau_1 - K\tau_1 + K\tau_2} \right]$$

$$= 10 \log \left[\frac{\dfrac{1}{\tau_1}}{\dfrac{1}{\tau_1 - K\tau_1 + K\tau_2}} \right]$$

$$= 10 \log \left[\frac{\tau_1 - K\tau_1 + K\tau_2}{\tau_1} \right]$$

$$= 10 \log \left[1 - K + \frac{K\tau_2}{K_1} \right]$$

Now, since $TL_1 - TL_2 = 10 \log \dfrac{\tau_2}{\tau_1}$,

$$\frac{TL_1 - TL_2}{10} = \log \frac{\tau_2}{\tau_1}$$

$$10^{\frac{TL_1 - TL_2}{10}} = \frac{\tau_2}{\tau_1}$$

Therefore,

$$TL_1 - TL_c = 10 \log \left[1 - K + K\, 10^{\frac{TL_1 - TL_2}{10}} \right] \qquad (6.4)$$

In this equation, TL_1 is the transmission loss of the part of the panel with the larger TL; TL_2 is the transmission loss of the smaller TL section; and TL_c is the transmission loss of the composite wall. Therefore, $TL_1 - TL_c$ is the number of decibels to be subtracted from the high TL section to obtain the

TL of the composite wall. The percent of the panel area with the lower transmission loss is *K*.

EXAMPLE 6.2   A 4-in. thick brick wall has a transmission loss of 45 dB. A window with $\frac{1}{4}$-in. thick glass is cut into the wall, and the window area is 10 percent of the wall area. If the transmission loss of the glass at 125 Hz is 15 dB, what is the transmission loss of the wall and window together?

$$TL_1 - TL_2 = 45 - 15 = 30$$

and

$$K = 10\% = 0.10$$

Therefore,

$$TL_1 - TL_c = 10 \log (1 - 0.10 + 0.10 \times 10^3)$$
$$= 10 \log (1 - 0.10 + 100.)$$
$$= 20 \text{ dB}$$

and the transmission loss of the composite wall is $45 - 20 = 25$ dB.

Equation (6.4) can be shown in the form of curves, as in Fig. 6.4, for convenience in estimating transmission loss in such cases.

It is obvious that small leaks in a barrier or enclosure can destroy the effectiveness of the enclosure. For example, a hole or crack with an area of only 1 percent of the barrier area can reduce transmission loss from 45 to only 20 dB. The *TL* of the opening is zero, and Fig. 6.4 shows that in this case

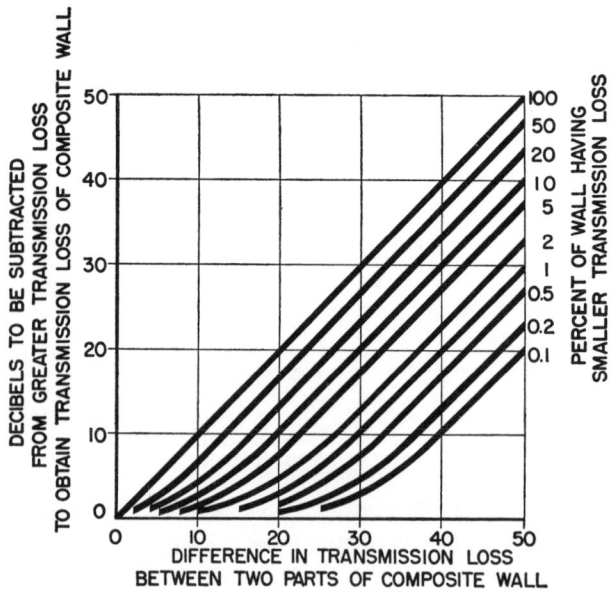

Figure 6.4

25 dB must be subtracted from the original 45 dB. It can also be seen that a leak in a high-quality enclosure is more damaging than it would be in a poorer quality one, since they both end up with nearly the same overall transmission loss.

For example, in Fig. 6.4 a 50-dB enclosure with a leak of 1 percent of the area has a *TL* of 20 dB. A 35-dB enclosure with the same leakage area has the same final *TL* of 20 dB.

6.4 Noise Reduction

The effectiveness of a wall or partition involves both the transmission loss of the wall and the acoustic properties of the receiving room. That is, a noise source on one side of an acoustic barrier produces a relatively high noise in a room on the other side of the barrier if there is only a small amount of absorption in the receiving room. If larger amounts of absorption are present in the receiving room, the sound level is somewhat less.

Transmission loss is measured in a laboratory by making the test panel act as the common wall between two rooms. The difference in sound pressure levels between the source room and the receiving room is called the noise reduction (*NR*), and this is related to the transmission loss by the following equation:

$$TL = NR + 10 \log \frac{S_{\text{wall}}}{S_{\text{room}} \, \bar{\alpha}_{\text{room}}} \qquad (6.5)$$

where TL = the transmission loss of the wall, in decibels,

NR = the difference in sound pressure level between the source room and the receiving room, in decibels,

S_{wall} = the area of the common wall, in square feet, and

S_{room} = the area of walls, ceiling, and floor of the receiving room, in square feet, and

$\bar{\alpha}$ = the average absorption coefficient in the receiving room.

Transmission loss measured in the laboratory and calculated by equation (6.5) is usually higher than it would be if measured under field conditions. In the laboratory, care is taken to prevent flanking noise transmission and to make sure that noise generated in the source room is transmitted to the receiving room almost entirely through the test panel. In actual field measurements these conditions often are not met.

Solving equation (6.5) for *NR* and substituting for *TL* its equivalent from equation (6.1):

$$NR = TL - 10 \log \frac{S_{\text{wall}}}{S_{\text{room}} \, \bar{\alpha}_{\text{room}}}$$

$$TL = 10 \log \frac{1}{\tau}$$

$$NR = 10 \log \frac{1}{\tau_{\text{wall}}} - 10 \log \frac{S_{\text{wall}}}{S_{\text{room}} \, \bar{\alpha}_{\text{room}}}$$

$$= 10 \log \frac{S_{\text{room}} \, \bar{\alpha}_{\text{room}}}{S_{\text{wall}} \, \tau_{\text{wall}}}$$

$$= 10 \log \frac{\text{receiving room absorption, in sabins}}{\text{wall transmittance, in square feet}} \qquad (6.6)$$

This is sometimes called noise insulation factor.

EXAMPLE 6.3   A room 50-ft long, 30-ft wide, and 15-ft high is to be built as an engineering office next to a shop area where machinery noise is measured at 90 dB. The walls of the room are to be 4-in. brick, the ceiling is to be 4-in. painted concrete, the floor is to be 4-in. concrete, and there will be two $1\frac{3}{4}$-in. solid-core oak doors in the end walls, each 3-ft wide by 8-ft high. The walls will be treated with sound-absorbing material that has an absorption coefficient of 0.75. Calculate the sound pressure level inside the office when the machinery is operated outside.

PROCEDURE

1. Estimate transmission loss of ceiling, floor, walls, and doors from published tables.
2. Calculate transmission coefficients of various areas by equation (6.1).
3. Calculate transmittance of various areas.
4. Calculate total transmittance.
5. Calculate total room absorption.
6. Calculate noise reduction by equation (6.6).
7. Calculate SPL in office.

   First, from equation (6.1),

$$TL = 10 \log \frac{1}{\tau}$$

$$\frac{1}{\tau} = \text{antilog} \frac{TL}{10}$$

	Area, ft²	TL, dB	τ	τs
Ceiling	1500	50	0.0000100	0.0150
Floor	1500	50	0.0000100	0.0150
Side Walls	1500	45	0.0000316	0.0474
End Walls	852	45	0.0000316	0.0269
Two Doors	48	20	0.01	0.4800
Total Transmittance				0.5843

Total absorption in room from Table 4.1:

	Area, ft²	α	αs
Ceiling	1500	0.07	105.
Floor	1500	0.02	30
Side Walls	1500	0.75	1125.
End Walls	852	0.75	639
Two Doors	48	0.07	3.
Total Absorption, Sabins			1902.

Noise reduction:

$$NR = 10 \log \frac{1902}{0.5843}$$

$$= 10 \log 3260$$

$$= 10\,(3.51)$$

$$= 35.\ dB$$

Therefore, the sound pressure level inside the engineering office is 90 − 35 = 55 dB.

6.5   Acoustic Enclosures

When machinery noise must be reduced 20 dB or more, it is usually necessary to use complete enclosures. As discussed previously, the attenuation of an enclosure or barrier, at any particular frequency, depends on the stiffness, mass, damping, and resonances. In general, with practical machinery enclosures, the lowest resonant frequency is below the lowest sound frequency of interest. That is, the walls behave as though they were mass-controlled and vibrate back and forth as a unit. Some of the smaller ones operate in the stiffness-controlled region where the exciting frequencies are below the fundamental resonance of the enclosure. It is important, therefore, to design the enclosure for the proper frequency range.

It must be kept in mind that the actual decrease in noise produced by the enclosure depends on other things as well as the transmission loss of the enclosure material. Vibration resonances must be avoided, or their effects reduced by damping; structural and mechanical connections must not be permitted to short circuit the enclosure; and the enclosure must be sealed as well as possible to prevent acoustic leaks. In addition, the actual noise reduction depends on the acoustic properties of the room in which the enclosure is located. For this reason, published data on transmission loss of various materials should not be assumed to be the same as the noise reduction which will be obtained when using those materials in enclosures.

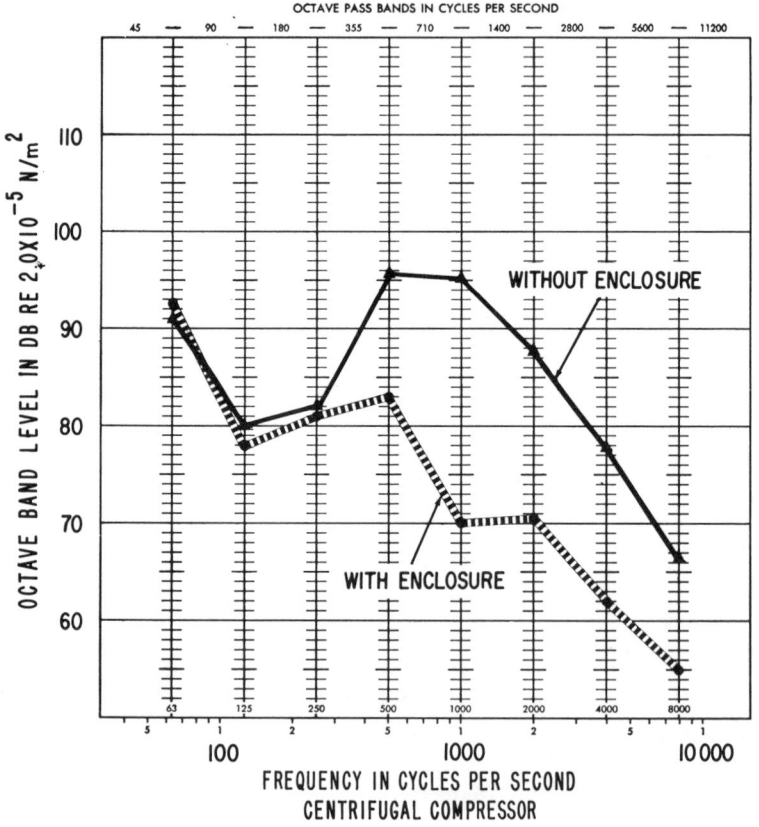

How materials are used in machinery enclosures is just as important as which materials are used.

In many applications acoustic enclosures provide an excellent means for reducing machinery noise to acceptable levels. The relationship between noise reduction and the acoustic properties of the receiving room is given by the following equation:

$$NR = TL - 10 \log \left[\frac{1}{4} + \frac{S_{\text{wall}}}{R_{\text{room}}} \right] \tag{6.7}$$

where NR = the noise reduction, or the difference in sound pressure level between the source room and the receiving room, in decibels,

S_{wall} = the area of the enclosure walls, in square feet, and

R_{room} = the room constant of the receiving room, in square feet.

EXAMPLE 6.4   A centrifugal air compressor produces 110 dB in the 1000-Hz octave band at a distance of 3 ft from its nearest major surface. The operator of another machine is 20 ft from the compressor.

An enclosure is to be installed over the compressor to reduce the sound pressure level in the 1000-Hz band to 85 dB at the operator's position.

The compressor is 6-ft long, 6-ft wide, and 4-ft high, and it is located in a room 75-ft long, 60-ft wide, and 25-ft high. The floor and ceiling of the room are concrete, and the walls are concrete block.

Calculate the required transmission loss of the enclosure.

(a) Calculate the average absorption coefficient in the room.

$$\bar{\alpha} = \frac{\alpha_1 S_1 + \alpha_2 S_2 + \alpha_3 S_3}{S_1 + S_2 + S_3} \tag{4.2}$$

	Surface Area, ft²	Absorption Coefficient	Absorption, Sabins
Floor	4500	0.02	90
Walls	6750	0.29	1958.
Ceiling	4500	0.02	90
	15750		2138

$$\bar{\alpha} = \frac{2138}{15750} = 0.135$$

(b) Calculate the room constant.

$$R = \frac{S\bar{\alpha}}{1 - \bar{\alpha}} \qquad (4.4)$$

$$R = \frac{15750 \times 0.135}{1 - 0.135}$$

$$R = 2458. \text{ ft}^2$$

(c) Calculate the sound power level in the 1000-Hz octave band assuming that the acoustic center of the compressor is 3 ft inside the compressor. That is, the sound pressure level of 110 dB would be at a distance of 6 ft from the acoustic center of the machine.

$$PWL = SPL - 10 \log \left[\frac{1}{2\pi r^2} + \frac{4}{R} \right] - 10.5 \qquad (4.19)$$

$$= 110 - 10 \log \left[\frac{1}{2\pi \times 36} + \frac{4}{2458} \right] - 10.5$$

$$= 110 - 10 \log 0.006 - 10.5$$

$$= 110 - 10 [-2.22] - 10.5$$

$$= 110 + 22.2 - 10.5$$

$$= 122. \text{ dB re } 10^{-12} \text{ W}$$

(d) Calculate the SPL at the operator's position.

$$SPL = PWL + 10 \log \left[\frac{1}{2\pi r^2} + \frac{4}{R} \right] + 10.5 \qquad (4.18)$$

$$= 122 + 10 \log \left[\frac{1}{2\pi \times 529} + \frac{4}{2458} \right] + 10.5$$

$$= 122 + 10 \log [0.00192] + 10.5$$

$$= 122 + 10[-2.72] + 10.5$$

$$= 105 \text{ dB}$$

(e) Assume that the enclosure dimensions are 10-ft long, 10-ft wide, and 6-ft high, and that the average absorption coefficient of the inside of the enclosure is 0.75. The room constant for the enclosure is then

$$R = \frac{440 \times 0.75}{1 - 0.75} = 1320 \text{ ft}^2$$

(f) Calculate the *SPL* at the inner surface of the enclosure, 5 ft from the acoustic center of the compressor.

$$SPL = 122 + 10 \log \left[\frac{1}{2\pi \times 25} + \frac{4}{1320} \right] + 10.5$$

$$= 122 + 10 \log [-2.045] + 10.5$$

$$= 112 \text{ dB}$$

This sound pressure level is higher than the original 110 dB since it is 2 ft from the casing instead of 3 ft.

(g) Calculate the decrease in sound pressure level due to distance when moving from the location 2 ft from the casing (5 ft from acoustic center) to 20 ft from the casing (23 ft from acoustic center), and with no enclosure.

$$dB_{23} = dB_5 - 20 \log \frac{23}{5}$$

$$= 112 - 20 \log 4.6$$

$$= 112 - 20 (0.66)$$

$$= 99. \text{ dB}$$

(h) The required noise reduction for the enclosure is, therefore,

$$99 \text{ dB} - 85 \text{ dB} = 14 \text{ dB}$$

(i) The transmission loss of the enclosure walls must be found as follows:

$$NR = TL - 10 \log \left[\frac{1}{4} + \frac{S_{\text{wall}}}{R_{\text{room}}} \right] \tag{6.7}$$

The effective radiating area of the enclosure walls is 340 ft^2.

$$TL = 14 + 10 \log \left[\frac{1}{4} + \frac{340}{2458} \right]$$

$$TL = 14 + 10 \log 0.388$$

$$= 14 + 10[-0.41]$$

$$= 10 \text{ dB}$$

To be on the safe side, add 5 dB to the calculated *TL* since errors may have been made in the assumed values for absorption coefficients. Therefore, the required *TL* = $10 + 5 = 15$ dB.

6.5.1 Estimated Noise Reduction of Enclosure. When a noisy machine is covered by an enclosure with hard, reflective walls, the sound pressure level inside may be higher than it would be without the enclosure. The radiated sound is reflected from wall to wall, and it builds up until the rate of increase due to the reflections balances the rate of dissipation through the enclosure.

There would not be an increase in sound pressure level if sound-absorbing material were present inside to absorb the sound instead of reflect it. Equation (4.3) shows that reverberant sound decreases 3 dB for each doubling of the absorption.

Although porous, sound-absorbing material alone produces some transmission loss at very high frequencies, it provides practically no isolation in the usual machinery enclosure. Transmission loss must be provided by the dense, impervious walls of the enclosure itself.

An approximate method for estimating the noise reduction of an enclosure is

$$NR = 10 \log \left[1 + \frac{\bar{\alpha}}{\bar{\tau}} \right] \tag{6.8}$$

where NR = the noise reduction, in decibels,
    $\bar{\alpha}$ = the average absorption coefficient of the inside of the enclosure, and
    $\bar{\tau}$ = the average transmission coefficient of the enclosure.

EXAMPLE 6.5   A machinery enclosure is 8-ft long, 6-ft wide, and 5-ft high. It is lined with a material that has an absorption coefficient of 0.85 at 1000-Hz. At that same frequency the enclosure material is stated by the manufacturer to have a transmission loss of 40 dB. The machine is mounted on a concrete floor that has an absorption coefficient of 0.02. Calculate the expected noise reduction at 1000-Hz.

(a) The average absorption coefficient is

$$\bar{\alpha} = \frac{\alpha_1 S_1 + \alpha_2 S_2}{S_1 + S_2} \tag{4.2}$$

$$= \frac{(0.85 \times 188) + (0.02 \times 48)}{188 + 48}$$

$$= \frac{160}{236} = 0.683$$

(b) The average transmission coefficient is found from the transmission loss:

$$TL = 10 \log \frac{1}{\tau} \qquad (6.1)$$

$$\frac{1}{\tau} = \text{antilog} \frac{TL}{10}$$

$$= \text{antilog} \frac{40}{10}$$

$$= 10^4$$

Therefore,

$$\tau = 10^{-4} = 0.0001$$

(c) The expected noise reduction is then

$$NR = 10 \log \left[1 + \frac{0.683}{0.0001} \right]$$

$$10 \log (1 + 6830)$$

$$= 38 \text{ dB}$$

If the average absorption coefficient of the inside of the enclosure had been 0.02, the same as the concrete floor, the noise reduction would have been much less. That is,

$$NR = 10 \log \left[1 + \frac{0.02}{0.0001} \right]$$

$$= 10 \log (1 + 200)$$

$$= 23. \text{ dB}$$

Equation (6.8) shows that in the theoretical case where there is no sound absorption there is no noise reduction.

6.6 Double Walls

A 4-in. thick brick wall has a transmission loss of about 45 dB. An 8-in. thick brick wall, with twice as much weight, has a transmission loss of about 50 dB. After a certain point has been reached it is found to be impractical to try to obtain higher isolation values simply by doubling the weight, since both the weight and the cost become excessive, and only a 5-dB improvement is gained for each doubling of weight.

An increase can be obtained, however, by using double-wall construction. That is, two 4-in. thick walls separated by an air space are better than one 8-in. wall. However, noise radiated by the first panel can excite vibration of the second one and cause it to radiate noise. If there are any mechanical connections between the two panels, vibration of one directly couples to the other, and much of the benefit of double-wall construction is lost.

There is another factor that can reduce the effectiveness of double-wall construction. Each of the walls represents a mass, and the air space between them acts as a spring. This mass-spring-mass combination has a series of resonances that greatly reduce the transmission loss at the corresponding frequencies. The effect of the resonances can be reduced by adding sound-absorbing material in the space between the panels.

6.7   Acoustic Shields

A barrier or wall can be used to provide acoustic shielding in cases where complete covering of a machine or other noise source would interfere with its operation. A theoretical procedure, based on Fresnel diffraction of a wave from a line source parallel to an infinitely long edge, was presented by R. O. Fehr and R. J. Wells in 1951. Several modifications to the calculations have been made since then, based on field measurements, but in most cases the results actually obtained are only approximately equal to the calculated values.

An equation that is frequently used with fairly good accuracy is given below:

$$\text{reduction in dB} = 10 \log \frac{20H^2}{\lambda R} \qquad (6.9)$$

where H = the height of the barrier, in feet,
R = the distance from source to wall, in feet, and
λ = the wavelength of sound at the particular frequency, in feet.

The equation should be used only when the following conditions exist:

1. The distance from the wall to the receiver is much greater than the distance from the wall to the source.
2. The distance from the wall to the source is greater than the height of the wall.
3. The distance from the source to the ends of the wall is at least twice the distance R, as shown in Fig. 6.5.

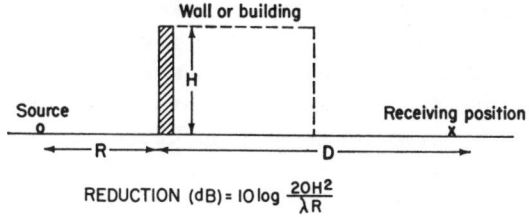

$$\text{REDUCTION (dB)} = 10 \log \frac{20H^2}{\lambda R}$$

Figure 6.5

EXAMPLE 6.6 An acoustic barrier is 40-ft high. The distance from a machine to the barrier is 200 ft. Calculate the noise reduction at 2000 Hz.

The wavelength of sound at 2000 Hz is calculated by equation (2.2).

$$\lambda = \frac{c}{f}$$

$$= \frac{1128}{2000} = 0.564 \text{ ft}$$

From equation (6.9),

$$\text{reduction in dB} = 10 \log \frac{20H^2}{\lambda R}$$

$$= 10 \log \frac{20 \times 40^2}{0.56 \times 200}$$

$$= 10 \log 286$$

$$= 25. \text{ dB}$$

Note that the noise reduction is much less at lower frequencies. At 125 Hz the calculated reduction is only 12 dB.

6.8   Vibration Isolation

Isolating a vibrating part from the rest of the structure can, in many instances, substantially reduce airborne sound. A vibration isolator, in its simplest form, is some type of resilient support. The purpose of the isolator may be to reduce the magnitude of force transmitted from a vibrating machine or part of a machine to its supporting structure. Conversely, its purpose may be to reduce the amplitude of motion transmitted from a vibrating support to a part of the system that is radiating noise due to its vibration.

Vibration isolators can be in the form of steel springs, cork, felt, rubber, plastics, or dense fiberglass. Steel springs can be calculated quite accurately and can do an excellent job of vibration isolation. However, they also can have resonances, and high frequency vibrations can travel through them readily, even though they are effectively isolating the lower frequencies. For this reason, springs usually are used in conjunction with elastomers or similar materials. Elastomers, plastics, and materials of this type have high internal damping and do not perform well below about 15 Hz. However, this is below the audible range and, therefore, does not limit their use in any way for effective sound control.

The noise reduction that can be obtained by installing an isolator depends on the characteristics of the isolator and the associated mechanical structure. For example, the attenuation that can be obtained by spring isolators depends not only on the spring constant, or spring stiffness (the force necessary to stretch or compress the spring one unit of length), but also on the mass load on the spring, the mass and stiffness of the foundation, and the type of excitation.

If the foundation is very massive and rigid, and if the mounted machine vibrates at constant amplitude, the reduction in force on the foundation is independent of frequency. If the machine vibrates at a constant force, the reduction in force depends on the ratio of the exciting frequency to the natural frequency of the system.

When a vibrating machine is mounted on an isolator, the ratio of the force applied to the isolator by the machine, to the force transmitted by the isolator, to the foundation is called the "transmissibility." That is,

$$\text{transmissibility} = \frac{\text{transmitted force}}{\text{impressed force}}$$

Under ideal conditions this ratio would be zero. In practice, the objective is to make it as small as possible. This can be done by designing the system so that the natural frequency of the mounted machine is very low compared to the frequency of the exciting force.

It can be shown that if no damping is present, the transmissibility can be expressed by the following equation:

$$T = \frac{1}{1 - (\omega/\omega_n)^2} \tag{6.10}$$

where $T =$ the transmissibility, expressed as a fraction

$\omega =$ the circular frequency of the exciting force, in radians per second, and

$\omega_n =$ the circular frequency of the mounted system, in radians per second.

When $\omega/\omega_n = 0$, the transmissibility equals 1.0. That is, there is no benefit obtained from the isolator.

If ω/ω_n is greater than zero, but less than 1.41, the isolator actually increases the magnitude of the transmitted force. This is called the "region of amplification." In fact, when ω/ω_n equals 1.0, the theoretical amplitude of the transmitted force goes to infinity, since this is the point where the frequency of the disturbing force equals the system natural frequency.

Equation (6.10) indicates that the transmissibility becomes negative when ω/ω_n is greater than 1.0. The negative number is simply due to the phase relation between force and motion, and it can be disregarded when considering only the amount of transmitted force.

Since vibration isolation is achieved only when ω/ω_n is greater than 1.41, the equation for transmissibility can be written so that T is positive:

$$T = \frac{1}{(\omega/\omega_n)^2 - 1}$$

The circular frequency is equal to 2π times the frequency in cycles per second. That is,

$$\omega = 2\pi f$$

By substituting this in the equation for transmissibility,

$$T = \frac{1}{(2\pi f/2\pi f_n)^2 - 1} = \frac{1}{(f/f_n)^2 - 1} \tag{6.11}$$

The static deflection of a spring when stretched or compressed by a weight is related to its natural frequency by the following equation:

$$f_n = 3.14\sqrt{\frac{1}{d}} \tag{6.12}$$

where $f_n =$ the natural frequency, in hertz, and
$d =$ the deflection, in inches.

By substituting the value of f_n in equation (6.11) for transmissibility,

$$T = \frac{1}{(f^2 d/3.14^2) - 1}$$

$$\frac{f^2 d}{3.14^2} - 1 = \frac{1}{T}$$

$$\frac{f^2 d}{3.14^2} = \frac{1}{T} + 1$$

$$d = \left(\frac{3.14}{f}\right)^2 \left(\frac{1}{T} + 1\right) \tag{6.13}$$

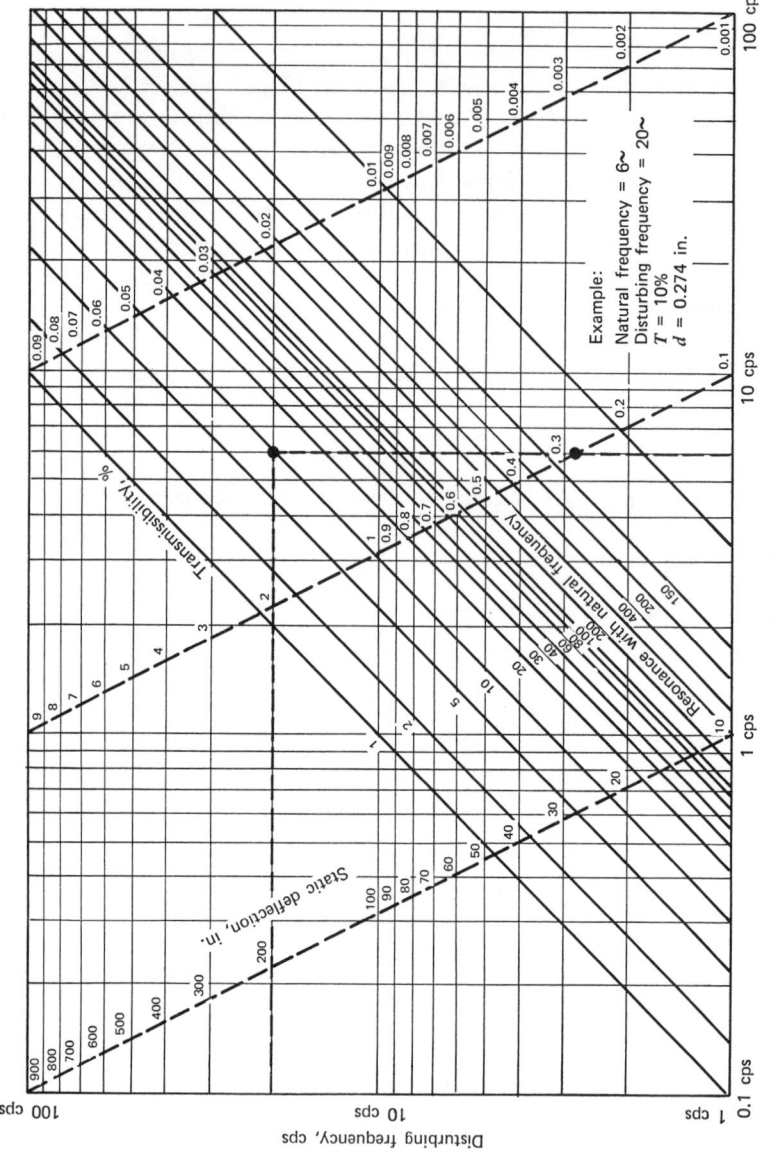

Figure 6.6 Transmissibility of flexible mountings. $T = 1/(\omega/\omega_n)^2 - 1$.

This shows that the transmissibility can be determined from the deflection of the isolator due to its supported load.

EXAMPLE 6.7 A machine rotates at 1200 RPM. Calculate the natural frequency of an isolator that limits the transmissibility to 10 percent.

(a) The exciting frequency, in this case the running speed, is

$$\frac{1200}{60} = 20 \text{ Hz}$$

(b)
$$T = \frac{1}{(f/f_n)^2 - 1} \tag{6.11}$$

$$0.10 = \frac{1}{(20/f_n)^2 - 1}$$

$$\left(\frac{20}{f_n}\right)^2 = \frac{1}{0.10} + 1 = 11$$

$$f_n = \frac{20}{\sqrt{11}} = 6. \text{ Hz}$$

EXAMPLE 6.8 A machine rotates at 1200 RPM. Calculate the required deflection of a spring isolator that limits the transmissibility to 10 percent. The exciting frequency is 20 Hz, the same as it was in Example 6.7.

$$d = \left(\frac{3.14}{f}\right)^2 \left(\frac{1}{T} + 1\right) \tag{6.13}$$

$$d = \left(\frac{3.14}{20}\right)^2 \left(\frac{1}{0.10} + 1\right)$$

$$= (0.0247)(11)$$

$$= 0.27 \text{ in.}$$

Equations (6.11) and (6.13) can be plotted, as shown in Fig. 6.6, for convenience in selecting isolator natural frequencies or deflections.

The answer to Example 6.7 can be found by drawing a vertical line from the intersection point of the 20-Hz disturbing frequency line and the 10-percent transmissibility diagonal. This shows the isolator natural frequency to be 6 Hz.

Similarly, in Example 6.8 the same vertical line as in Example 6.7 intersects the deflection line at 0.27 in. As shown, the transmissibility diagonals have positive slopes, whereas the deflection diagonals have negative slopes.

In general, it is not possible in the design stage to make an accurate calculation of the decrease in airborne noise to be obtained by vibration isolation, but it is always good practice to incorporate such isolation in every low-noise design. In some cases it is absolutely necessary.

For critical applications, the natural frequency of the isolator should be about one-tenth to one-sixth of the disturbing frequency. That is, the transmissibility should be between 1 and 3 percent. For less critical conditions, the natural frequency of the isolator should be about one-sixth to one-third of the driving frequency, with transmissibility between 3 and 12 percent.

In many installations there is no need to bolt isolation mounts to the foundation because the resulting smooth operation of the mounted machine and its weight are enough to prevent movement. When restraints are necessary or desirable for one reason or another, it is extremely important to keep the hold-down bolts or any other structural member from short-circuiting the vibration isolators.

Electrical conduit should be long and flexible. Piping should be flexible or supported on resilient mounts. Limit stops must not be in contact, and all objects under vibration-isolated equipment should be removed. An otherwise excellent job of vibration isolation can be made entirely ineffective by such objects as bolts, nuts, welding rods, and excess grout under the mounted machine.

6.9   Vibration Damping

Complex mechanical systems have many resonant frequencies, and whenever an exciting frequency is coincident with one of the resonant frequencies, the amplitude of vibration is limited only by the amount of damping in the system. If the exciting force is wide band, several resonant vibrations can occur simultaneously, thereby compounding the problem.

The transmissibility curves in Fig. 6.7 show the results of increasing the amount of damping in a vibrating system. At resonance a complex mechanical system can be considered to be a mass supported by a spring and a damper. In Fig. 6.7 the ordinate is shown as a ratio of the vibration amplitude to the deflection of the spring due to the weight of the supported mass. The abscissa is the ratio of the exciting frequency to the system natural frequency.

When the exciting frequency is less than the resonant frequency, the vibration amplitude is greater than the static deflection. This is called the "region of amplification," and it can be seen that in this region spring mounting does not help at all; in fact, it makes the problem worse.

If the exciting frequency is greater than 1.41 times the resonant frequency, the vibration amplitude is less than the static deflection. This is the "region of isolation."

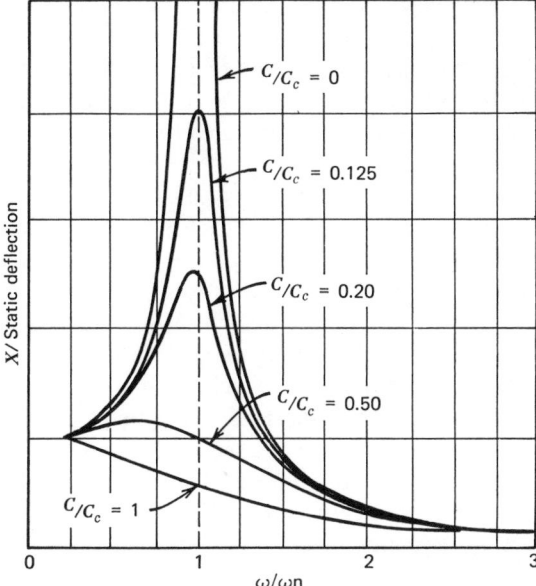

Figure 6.7

When the exciting frequency equals the system natural frequency, the amplitude goes to infinity if no damping is present. Figure 6.7 shows how damping decreases the vibration amplitude at resonance. It is convenient to express the amount of damping as a percent of critical damping. Critical damping is the amount of damping that must be added to a system so that it does not oscillate when the mass is displaced and then released.

It can be seen, therefore, that damping is one of the most important factors in noise and vibration control.

6.9.1 Three Types of Damping. There are three kinds of damping. (a) Viscous damping is the type that is produced by viscous resistance in a fluid such as a dash pot. The damping force is proportional to velocity. (b) Dry friction, or Coulomb damping, produces a constant damping force, independent of displacement and velocity. The damping force is produced by dry surfaces rubbing together, and it is opposite in direction to that of the velocity. (c) Hysteresis damping, also called material damping, produces a force that is in phase with the velocity, but is proportional to displacement. This is the type of damping found in solid materials, such as elastomers, widely used in sound control.

6.9.2   Free-Layer Damping.   A large amount of noise radiated from machine parts comes from vibration of large areas or panels. These parts can be integral parts of the machine or attachments to the machine. They can be flat or curved, and vibration can be caused by either mechanical or acoustic excitation. The radiated noise is a maximum when the parts are vibrating in resonance.

When the excitation is mechanical, vibration isolation may be all that is needed. In other instances, the resonant response can be reduced by bonding a layer of energy dissipating polymeric material to the structure. When the structure bends, the damping material is placed alternately in tension and compression, thus dissipating the energy as heat.

This extensional or free-layer damping is remarkably effective in reducing resonant vibration and noise in relatively thin, lightweight structures such as panels. It becomes less effective as the structure stiffness increases because of the excessive increase in thickness of the required damping layer.

In a vibrating structure, the amount of energy dissipated is a function of the amount of energy necessary to deflect the structure, compared to that required to deflect the damping material. If 99 percent of the vibration energy is required to deflect the structure, and 1 percent is required to deflect the damping layer, then only 1 percent of the vibration energy is dissipated.

6.9.3   Constrained-Layer Damping.   Resonant vibration amplitude in heavier structures can be controlled effectively by the application of constrained-layer damping. In this method, a relatively thin layer of viscoelastic damping material is constrained between the structure and a stiff cover plate. Vibration energy is removed from the system by the shear motion of the damping layer.

6.10   Mufflers

Silencers or mufflers are usually divided into two categories: absorptive or dissipative and reactive or reflective. Actually, all mufflers accomplish noise reduction by combining both effects, to a certain extent, so that the distinction is really arbitrary.

The performance of a muffler can be described in various ways. Unfortunately not everyone uses the same terminology, but, in general, the following definitions apply:

Insertion loss is defined as the difference between two sound pressure levels measured at the same point in space before and after a muffler is inserted in the system.

Dynamic insertion loss is the same as insertion loss, except that it is measured when the muffler is operating under rated flow conditions. Therefore, the dynamic insertion loss is of more interest than ratings based on no-flow conditions.

Transmission loss is defined as the ratio of sound power incident on the muffler to the sound power transmitted by the muffler. It cannot be measured directly, and it is difficult to calculate analytically. For these reasons the transmission loss of a muffler has little practical application.

Attenuation is used to describe the decrease in sound power as sound waves travel through the muffler. It does not convey information about how a muffler performs in a system.

Noise reduction is defined as the difference between sound pressure levels measured at the inlet of a muffler and those at the outlet.

6.10.1 Absorptive Mufflers.

Absorptive mufflers have relatively wide-band noise-reduction characteristics, and they are usually applied to noise control problems associated with continuous spectra, such as fans, centrifugal compressors, jet engines, and gas turbines. They are also used in cases where a narrow-band noise predominates, but the frequency varies because of a wide range of operating speeds.

A variety of sound-absorbing materials are used in many different configurations, determined by the level of the unsilenced noise and its frequency content, the type of gas being used, the allowable pressure drop through the silencer, the gas velocity, gas temperature and pressure, and the noise criterion to be met.

Fiberglass or mineral wool with density approximately 0.5 to 6.0 lb/ft³ is frequently used in absorptive silencers. These materials are relatively inexpensive and have good sound-absorbing characteristics. They operate on the principle that sound energy causes the material fibers to move, converting the sound energy into mechanical vibration and heat. The fibers do not become very warm since the sound energy is actually quite low, even at fairly high decibel levels.

Lined Ducts. The simplest kind of absorptive muffler is a lined duct, where the absorbing material is either added to the inside of the duct walls or the duct walls themselves are made of sound-absorbing material. The attenuation depends on the duct length, thickness of the lining, area of the air passage, type of absorbing material, and frequency of the sound passing through.

Parallel Baffle Mufflers. A common type of absorptive muffler is that which uses parallel or annular baffles, as shown in Figs. 6.8 and 6.9. The acoustical performance of such mufflers is a function of the length, thickness of the baffle sections, space between the baffles, and absorption coefficient of the material used in the baffles.

DISSIPATIVE MUFFLER

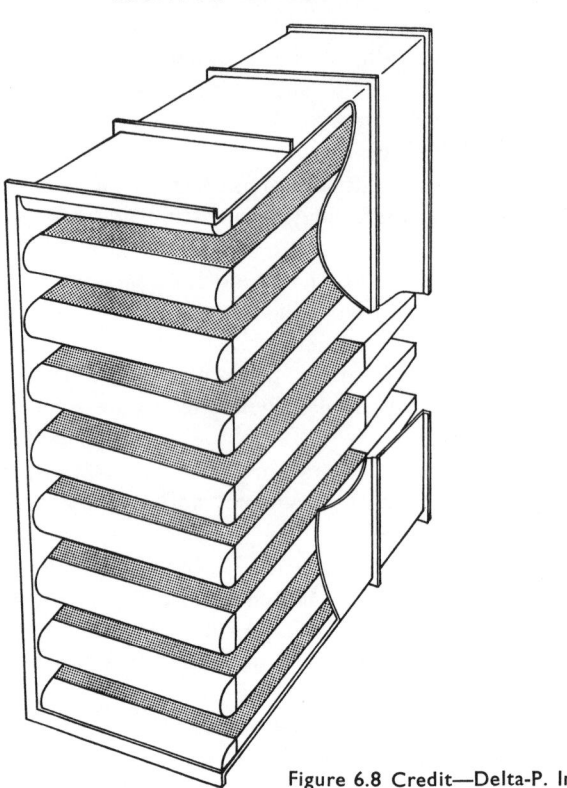

Figure 6.8 Credit—Delta-P. Inc.

The attenuation that can be obtained with parallel baffle-type silencers is approximately 3 to 9 dB/ft. Higher frequencies are attenuated more than lower frequencies, but improved low-frequency performance can be obtained by increasing the thickness and density of the baffles.

Figure 6.10 shows how the attenuation varies with baffle spacing. Figure 6.11 shows a typical performance curve illustrating how noise reduction varies with frequency.

Lined Bends.   When sound travels around a bend with an absorptive lining, the attenuation is greater than it would be in a straight section of duct. Greatest attenuation is obtained at the higher frequencies, but the actual amount is difficult to calculate analytically, and experimental data are not very consistent, mainly because the performance depends on the amount of scattering produced by the bend. Furthermore, measured sound levels depend

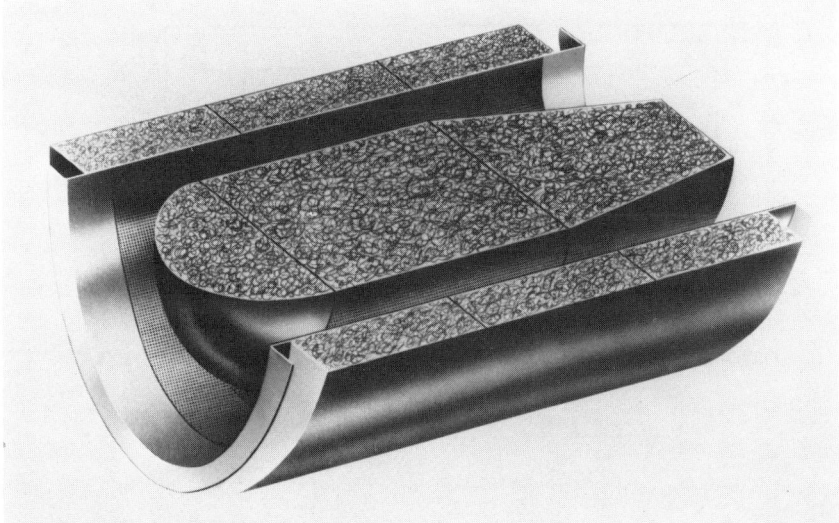

Figure 6.9 Credit—Industrial Acoustics.

on where microphones are located with respect to the bend, and changes in location can produce substantially different results.

Plenum Chambers. A plenum chamber is a chamber with a large volume and large cross section between two smaller ducts that are located at opposite ends and usually offset to minimize the direct transmission of sound. Figure 6.12 shows the usual arrangement. The inside of the plenum chamber is lined

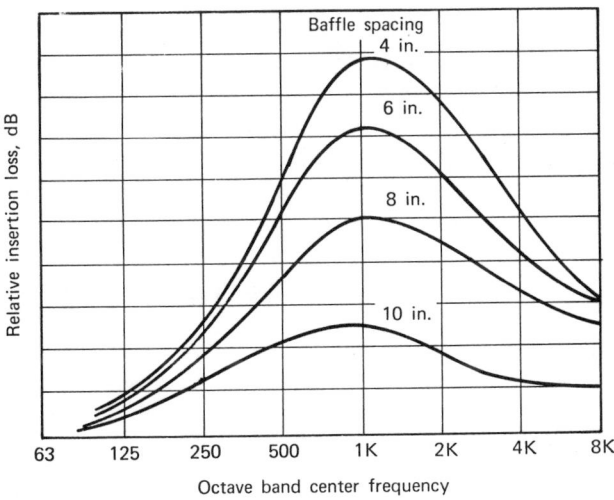

Figure 6.10

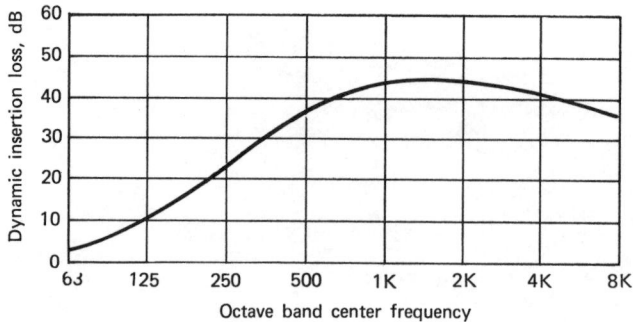

Figure 6.11 *Credit Industrial Acoustics??*

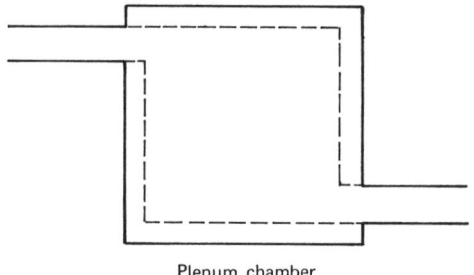

Plenum chamber

Figure 6.12

with sound-absorbing material that absorbs most of the sound energy as it is reflected back and forth from wall to wall.

The overall performance of the chamber is improved by increasing the ratio of plenum cross-section area to that of the inlet and outlet, and low-frequency performance can be helped by increasing the thickness of the sound-absorbing lining.

If a baffle is added to the chamber between the inlet and outlet, the performance is improved further since this forces the sound to be in contact with the sound-absorbing material over a greater distance.

6.10.2 Reactive Mufflers. Reactive mufflers have a characteristic acoustic performance that does not depend to any great extent on the presence of sound-absorbing material, but utilizes the reflection characteristics and attenuating properties of conical connectors, expansion chambers, side branch resonators, tail pipes, and so on, to accomplish sound reduction. Figure 6.13 shows a typical reactive muffler.

Expansion chambers operate most efficiently in applications involving discrete frequencies rather than broad band noise. The length of the chamber

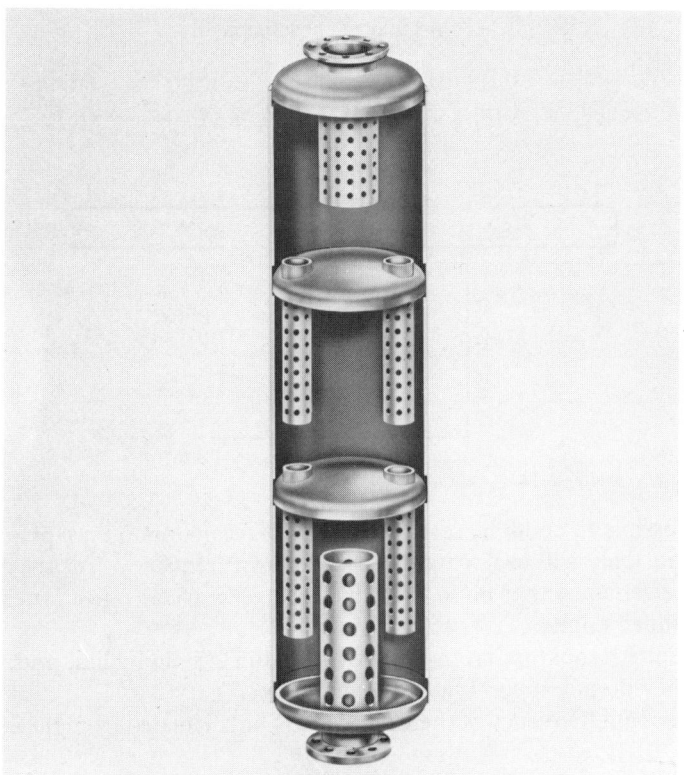

Figure 6.13   1340:1 BEO snubbers (courtesy of Burgess Industries).

is adjusted so that reflected waves cancel the incident waves, and since wavelength depends on frequency, expansion chambers should be tuned to some particular frequency. When a number of discrete frequencies must be attenuated, several expansion chambers can be placed in series, each tuned to a particular wavelength.

6.10.3   Helmholtz Resonator.   A Helmholtz resonator is a vessel containing a volume of air that is connected to a noise source such as a piping system, as shown in Fig. 6.14. When a pure-tone sound wave is propagated along the conduit, the air in the vessel expands and contracts. By proper design of the area and length of the neck, and volume of the chamber, sound wave cancellation can be obtained, thereby reducing the tone. This type of resonator can be designed to produce maximum reduction over a very narrow frequency range, as shown in Fig. 6.14.

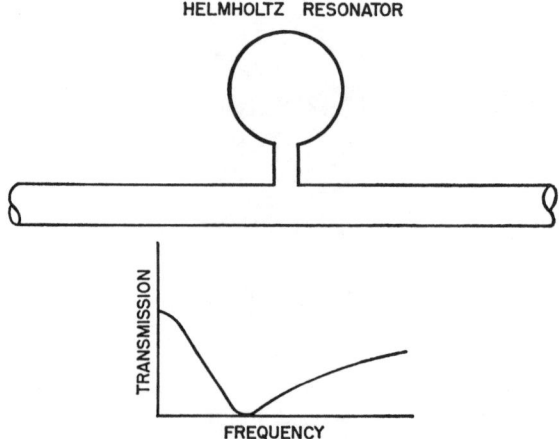

Figure 6.14

It is possible to combine several Helmholtz resonators on a piping system so that not only will each cancel out at its own frequency, but they can be made to overlap so that noise is attenuated over a wider range instead of at sharply tuned points.

Helmholtz resonators are normally located in side branches, and for this reason they do not affect flow in the main pipe.

The resonant frequency of these devices can be calculated from the equation

$$f_0 = \frac{C}{2\pi}\sqrt{\frac{A}{LV}} \text{ Hz} \tag{6.14}$$

where C = the speed of sound in the fluid, in feet per second,
A = the cross-sectional area of the neck, in square feet,
L = the length of the neck, in feet, and
V = the volume of the chamber, in cubic feet.

EXAMPLE 6.9   Design a Helmholtz resonator to reduce a 250-Hz noise component in an air pipe line.

(a)
$$f_0 = \frac{C}{2\pi}\sqrt{\frac{A}{LV}} \tag{6.14}$$

$$\sqrt{\frac{A}{LV}} = \frac{2\pi f_0}{C}$$

$$\frac{A}{LV} = \left[\frac{2\pi f_0}{C}\right]^2$$

$$V = \frac{A}{(2\pi f_0/C)^2 L} = \frac{C^2 A}{(2\pi f_0)^2 L}$$

(b) Let $C = 1128$ ft/sec $= 13550$ in./sec.

$$V = \frac{13550^2\,A}{(6.28 \times 250)^2 L} = \frac{74.5A}{L}$$

(c) Try $A = 0.25$ in.^{2} and $L = 0.50$ in.
Then

$$V = \frac{74.5 \times 0.25}{0.50} = 37.25 \text{ in.}^3$$

6.10.4 Quincke Tube.   The Quincke tube, illustrated in Fig. 6.15, is another type of reactive muffler. If the pipe is divided into two sections with one leg

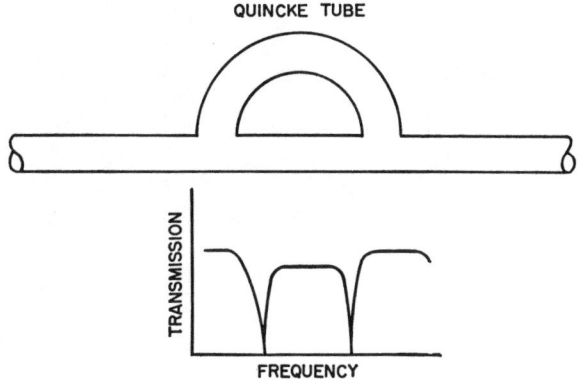

Figure 6.15

a half wavelength longer than the other, phase cancellation occurs since the two sound waves come together 180° out of phase. The two paths should be designed to carry equal amounts of flow.

EXAMPLE 6.10   Calculate the difference in path length of two sections of a Quincke tube in order to cancel a 100-Hz tone in a water pipe system.

(a) Assume that the speed of sound in water, for this case, is 4720 fps.

(b) The wavelength of a 100-Hz sound can be calculated by means of equation (2.2).

$$C = f\lambda \tag{2.2}$$

$$\lambda = \frac{C}{f} = \frac{4720}{100} = 47.20 \text{ ft}$$

(c)

$$\frac{\lambda}{2} = \frac{47.20}{2} = 23.60 \text{ ft}$$

(d) Therefore, if the loop section is 23.60-ft longer than the straight section, then total cancellation of the 100-Hz tone occurs. Cancellation also occurs at 300 and 500 Hz.

Cancellation is sharp, but slight changes in the speed of sound due to changes in temperature and pressure make the Quincke tube impractical in most cases.

6.11   Typical Gas Turbine Sound Control

Many city and local governments are establishing maximum permissible sound levels for various noise-sensitive zones. This means that industries planning new installations, or making changes in existing ones, must make sure that a noisy machine or group of machines complies with the noise control code. For this reason, the selection of adequate sound control equipment becomes an important part of the planning of new facilities.

Gas turbine-driven-compressor installations are typical examples. Many are located near residential areas, or where future residential areas are expected. Without proper sound control these installations are quite noisy and can be expected to provoke complaints from nearby communities.

In order to reduce noise levels to acceptable values, the same procedure must be followed as in all other noise control projects. The major noise sources must be identified and reduced. Sometimes several can be treated at the same time, but usually they must be tackled one by one.

The predominant sources of noise in gas turbine-compressor installations, in order of importance, are the gas turbine inlet, gas turbine exhaust, and gas turbine casing radiation. In general, the noise generated by the compressor is well below that produced by the gas turbine, and for that reason the compressor is not treated in some cases. For most critical applications, the compressor is enclosed along with the gas turbine.

Noise radiated by compressor piping can be quite high, but it can be reduced by adequate lagging.

Gas turbine inlet noise, the major component, is caused by interaction of rotor and stator blades. This high-level, high-frequency noise must be reduced first, and this can be done with an inlet silencer.

Exhaust noise generated by the combustion process produces high-level, low-frequency noise. It too can be reduced satisfactorily with a silencer.

Casing radiation, usually 10 to 15 dB below the combined inlet and exhaust noise, can be reduced to acceptable levels with an acoustic enclosure.

A typical problem may be one that requires the determination of the dynamic insertion loss of an inlet silencer, the dynamic insertion loss of an exhaust silencer, and the transmission loss of an acoustic enclosure in order to meet a particular boundary noise criterion.

EXAMPLE 6.11   Sound tests conducted on a gas turbine show that the octave band sound power levels radiated by the inlet, exhaust, and casing are as listed below:

Octave	63	125	250	500	1K	2K	4K	8K
Inlet PWL re 10^{-12} W	117	121	124	126	125	130	131	134
Exhaust PWL re 10^{-12} W	136	132	132	131	126	119	117	111
Casing PWL re 10^{-12} W	116	120	120	124	124	124	127	125

Calculate the dynamic insertion loss of inlet and exhaust silencers and the transmission loss of an acoustic enclosure for the turbine so that the following octave band sound pressure levels will not be exceeded at a distance of 500 ft from the machine.

Octave	63	125	250	500	1K	2K	4K	8K
dB at 500 ft	75	66	59	54	50	47	45	44

The octave band sound pressure levels at 500 ft can be calculated by equation (4.12):

$$SPL = PWL - 20 \log r + 2.5$$

The levels radiated by the inlet and exhaust depend also on the direction in which they are pointing. Equation (4.6) shows that the directivity index is equal to the amount by which the sound pressure level in any particular direction exceeds the average sound pressure level. Assume that the inlet and exhaust both point vertically upward and that the directivity values in this 90° direction are as listed in the calculation below:

	Inlet							
Octave	63	125	250	500	1K	2K	4K	8K
PWL re 10^{-12} W	117	121	124	126	125	130	131	134
Divergence to 500 ft ($SPL = PWL - 20$ $\log r + 2.5$)	65	69	72	74	73	78	79	82
Directivity	−3	−6	−8	−11	−14	−16	−16	−18
SPL at 500 ft	62	63	64	63	59	62	63	64
Design Citerion	75	66	59	54	50	47	45	44
Silencer Insertion Loss	(−13)	(−3)	5	9	9	15	18	20
Required Silencer Insertion Loss with Three Sources	(−8)	2	10	14	14	20	23	25

The resultant sound pressure levels at 500 ft are the sum of the inlet, exhaust, and casing contributions. This means that if each of these three meets the design criterion, the sum exceeds it. Assuming equal contributions from each, the total is increased by 10 log 3, or 5 dB [equation (2.11)]. Therefore, the required dynamic insertion loss of the silencer should be increased by 5 dB.

	Exhaust							
Octave	63	125	250	500	1K	2K	4K	8K
PWL re 10^{-12}	136	132	132	131	126	119	117	111
Divergence to 500 ft	84	80	80	79	74	67	65	59
Directivity	−1	−4	−7	−9	−12	−15	−18	−18
SPL at 500 ft	83	76	73	70	62	52	47	41
Design Criterion	75	66	59	54	50	47	45	44
Silencer Insertion Loss	8	10	14	16	12	5	2	(−3)
Required Silencer Insertion Loss with Three Sources	13	15	19	21	17	10	7	2

	Casing							
Octave	63	125	250	500	1K	2K	4K	8K
PWL re 10^{-12} W	116	120	120	124	124	124	127	125
Divergence to 500 ft	64	68	68	72	72	72	75	73
Directivity	0	0	0	0	0	0	0	0
SPL at 500 ft	64	68	68	72	72	72	75	73
Design Criterion	75	66	59	54	50	47	45	44
Enclosure Transmission Loss	(−11)	2	9	18	22	25	30	29
Required Enclosure Transmission Loss with Three Sources	(−6)	7	14	23	27	30	35	34

The acoustic enclosure should also include the compressor, and the compressor piping outside the enclosure should be lagged by covering it with 2 to 3 in. of fiberglass with a density of 3 to 6 lb/ft³ and an outer covering weighing approximately 1 lb/ft². This outer covering can be leaded vinyl, weighing about 0.87 lb/ft², or #24 gage sheet steel, or the equivalent weight of aluminum.

6.12 Sound Control Recommendations

A list of recommendations for machinery noise reduction follows:

1. Reduce horsepower. Noise is proportional to horsepower. Therefore, the machine should be matched to the job. Excess horsepower means excess noise.
2. Reduce speed. Slow-speed machinery is quieter than high-speed machinery.
3. Keep impeller tip speeds low. However, it is better to keep the RPM low and the impeller diameter large than to keep the RPM high and the impeller diameter small, even though the tip speeds are the same.
4. Improve dynamic balance. This decreases rotating forces, structure-borne sound, and the excitation of structural resonances.
5. Reduce the ratio of rotating masses to fixed masses.
6. Reduce mechanical run-out of shafts. This improves the initial static and dynamic balance.
7. Avoid structural resonances. These are often responsible for many unidentified components in the radiated sound. In addition to being excited by sinusoidal forcing frequencies, they can be excited by impacting parts and sliding and rubbing contacts.
8. Eliminate or reduce impacts. Either reduce the mass of impacting parts or their striking velocities.
9. Reduce peak acceleration. Reduce the rate of change of velocity of moving parts by utilizing the maximum time possible to produce the required velocity change, and by keeping the acceleration as nearly constant as possible over the available time period.
10. Improve lubrication. Inadequate lubrication is often the cause of bearing noise, structure-borne noise due to friction, and the excitation of structural resonances.
11. Maintain closer tolerances and clearances in bearings and moving parts.
12. Install bearings correctly. Improper installation accounts for approximately half of bearing noise problems.
13. Improve alignment. Improper alignment is a major source of noise and vibration.
14. Use center of gravity mounting whenever feasible. When supports are symmetrical with respect to the center of gravity, translational modes of vibration do not couple to rotational modes.
15. Maintain adequate separation between operating speeds and lateral and torsional resonant speeds.

16. Consider the shape of impeller vanes from an acoustic standpoint. Some configurations are noisier than others.
17. Keep the distance between impeller vanes and cutwater or diffuser vanes as large as possible. Close spacing is a major source of noise.
18. Select combinations of rotating and stationary vanes that are not likely to excite strong vibration and noise.
19. Design turning vanes properly. They are a source of self-generated noise.
20. Keep the areas of inlet passages as large as possible and their length as short as possible.
21. Remove or keep at a minimum any obstructions, bends, or abrupt changes in fluid passages.
22. Pay special attention to inlet design. This is extremely important in noise generation.
23. Item 22 applies also to the discharge, but the inlet is more important than the discharge from an acoustic standpoint.
24. Maintain gradual, not abrupt, transition from one area to the next in all fluid passages.
25. Reduce flow velocities in passages, pipes, and so on. Noise can be reduced substantially by reducing flow velocities.
26. Reduce jet velocities. Jet noise is proportional to the eighth power of the velocity.
28. Reduce large radiating areas. Surfaces radiating certain frequencies can often be divided into smaller areas with less radiating efficiency.
28. Disconnect possible sound radiating parts from other vibrating parts by installing vibration breaks to eliminate metal to metal contact.
29. Provide openings or air leaks in large radiating areas so that air can move through them. This reduces pressure build-up and decreases radiated noise.
30. Reduce clearances, piston weights, and connecting rod weights in reciprocating machinery to reduce piston impacts.
31. Apply additional sound control devices, such as inlet and discharge silencers and acoustic enclosures.
32. When acoustic enclosures are used make sure that all openings are sealed properly.
33. Install machinery on adequate mountings and foundations to reduce structure-borne sound and vibration.
34. Take advantage of all directivity effects whenever possible by directing inlet and discharge openings away from listeners or critical areas.
35. When a machine must meet a particular sound specification, purchase driving motors, turbines, gears, and auxiliary equipment that produce 3- to 5-dB lower sound levels than the machine alone. This insures that the combination is in compliance with the specification.

PART THREE

NOISE CRITERIA AND SPECIFICATIONS

CHAPTER 7

NOISE CONTROL
CRITERIA

7.1 Hearing Damage Criteria

It is a well-known fact that overexposure to high noise levels causes hearing damage. Not so well-known however is how much damage is caused by particular combinations of decibel level exposure and time.

A report by the National Academy of Science, National Research Council Committee on Hearing, Bioacoustics, and Biomechanics, entitled "Hazardous Exposure to Intermittent and Steady-State Noise" (*ASA Journal* Vol. 39, No. 3, March 1966) provided a method for the assessment of possible hearing damage resulting from such exposures, and it became the basis for selecting permissible sound levels in the Federal Walsh-Healey Public Contracts Act.

Figure 7.1 shows octave band data taken from the report. It can be seen that higher sound levels are permitted at the low-frequency end of the spectrum than at the high-frequency end because the ear is more sensitive to high frequencies than to low. Sounds must be at higher level at low frequencies to seem as loud as high-frequency sounds.

Analysis of the data in this detailed study, known as the CHABA report, and other investigations about hearing damage levels have revealed that a simpler approach can be made, based on *A*-weighted overall levels, instead of the more complicated octave band and third-octave band method. This dB*A* rating may be somewhat oversimplified, and may not be an exact statement of the problem, but it offers a method that can be understood and implemented by people in industry, and accomplishes its main purpose.

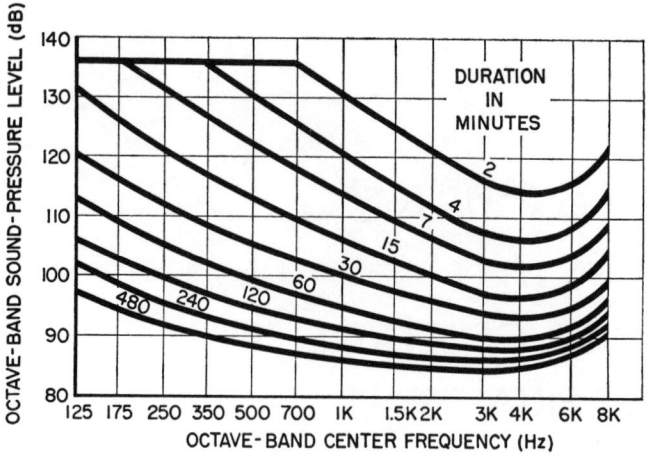

Figure 7.1

It can be seen in Fig. 7.1 that the shape of the hearing damage curves is the reverse of the A-weighted response curve in sound level meters.

The Occupational Safety and Health Act, Public Law 91-596, superseded the Walsh-Healey Act, but retained the same permissible sound exposure limits. These levels are listed in Table 7.1.

Potential hearing damage is a function of both decibel level and time of exposure. On an equal energy basis, this means that the permissible level could be increased 3 dBA each time exposure time is halved. The Occupational Safety and Health Act permits a 5-dBA increase instead of 3 dBA, when the

TABLE 7.1

Duration per Day	Sound Level, dBA
Hours	Slow Response
8	90
6	92
4	95
3	97
2	100
$1\frac{1}{2}$	102
1	105
$\frac{1}{2}$	110
$\frac{1}{4}$ or less	115

time is halved, to allow a reasonable correction for intermittency of the sound.

The act also states that in no case should a worker be exposed to more than 115 dBA, however short the exposure, nor to impulsive or impact noise in excess of 140 dBC, that is, when measured with the C-weighting network. This latter level requires special instrumentation, such as an impact noise analyzer or a cathode-ray oscilloscope. It cannot be read on an ordinary sound level meter.

Even at 90-dBA exposure for 8 hr/day, there is a risk of hearing damage to some people. The standard is considered to prevent disabling loss of hearing in 80 percent of those exposed.

The Occupational Safety and Health Act offers an alternate procedure for determining the equivalent A-weighted sound level from octave band sound pressure levels. The octave band data are plotted on a chart of equivalent sound level contours (Fig. 7.2). The point of highest penetration into the sound contours determines the A-weighted sound level. This dBA level is used to evaluate the exposure limit, even though it may differ slightly from the actual A-weighted sound level.

For example, if the octave band sound pressure levels from Example 3.3 are plotted on the contours, Fig. 7.2, the point of highest penetration is found

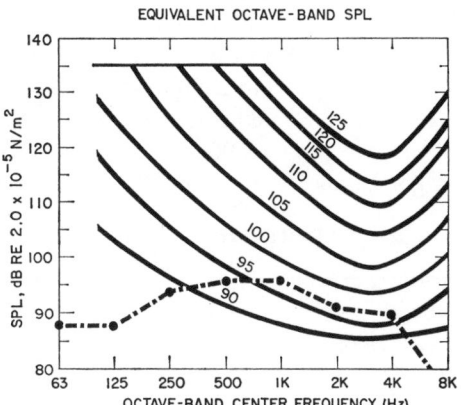

Figure 7.2

to be 97 dBA. This is 3-dBA less than the calculated value of 100 dBA. The octave band levels in Example 3.3 were the following:

Octave	63	125	250	500	1K	2K	4K	8K
SPL	88	88	94	96	96	92	89	76

When the daily exposure is composed of two or more periods of noise exposure of different levels, the combined effect should be considered, rather than the individual effect of each. That is, if the sum of the following fractions exceeds unity, then the mixed exposure is considered to exceed the limit value:

$$\frac{C_1}{T_1} + \frac{C_2}{T_2} + \cdots \frac{C_n}{T_n} > 1 \qquad (7.1)$$

where C_n = the total time of exposure at a specified noise level, and
T_n = the permissible time at that exposure.

EXAMPLE 7.1   The noise levels in a work area are found to be 95 dBA for 3 hr, 90 dBA for 4 hr, and 85 dBA for 1 hr. Calculate the combined exposure.

From Table 7.1, the permissible time for 95 dBA is 4 hr, for 90 dBA it is 8 hr, and for 85 dBA there is no limit. Therefore from equation (7.1) the combined exposure is

$$\frac{3}{4} + \frac{4}{8} + 0 = 1.25$$

This exceeds the permissible limit of 1.0.

7.2   Speech Interference Levels

The objective in a sound control program may be to avoid levels that interfere with speech communication. Speech interference level (*SIL*) is defined as the arithmetic average of the sound pressure levels in the three octave bands: 600 to 1200 Hz; 1200 to 2400 Hz; and 2400 to 4800 Hz.

In terms of the new, preferred octaves, it is called *PSIL*, and it is the arithmetic average of the levels in the 500-, 1000-, and 2000-Hz octaves. The resulting number is a good indication of how noise interferes with speech.

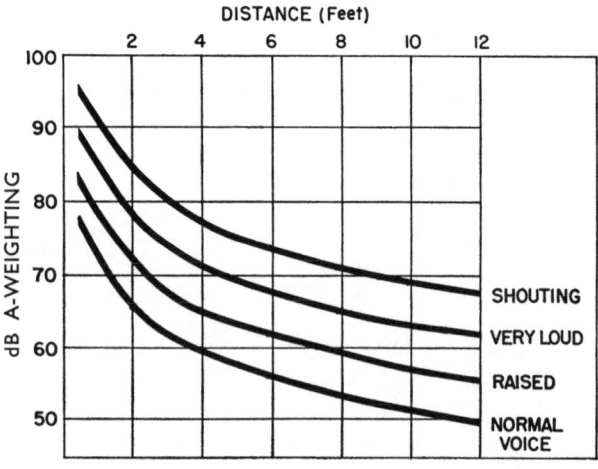

Figure 7.3

Speech interference levels can also be given in terms of A-weighted sound levels, or dBA, as read on a sound level meter. Figure 7.3 shows the levels that interfere with conversation at various distances.

No matter which criterion is used, it is important to remember that the ability to communicate depends on the intensity and frequency of the message as compared to the intensity and frequency of the background noise.

7.3 Annoyance Criteria

Preventing community complaints about noise is a growing problem for industry. Criteria to prevent annoyance are more restrictive than criteria to prevent hearing damage or speech interference. More factors enter into the problem.

Annoyance depends on the level of the offending noise compared to the preexisting background level, its absolute value, its frequency, how it varies with time, and whether it occurs during the day or night.

Whereas it is impossible to predict exactly the response from any particular neighborhood to any specific noise, a fairly reliable method has been developed by Rosenblith, Stevens, and the staff of Bolt, Beranek, and Newman.

Measured octave band sound pressure levels are plotted on a chart (Fig.

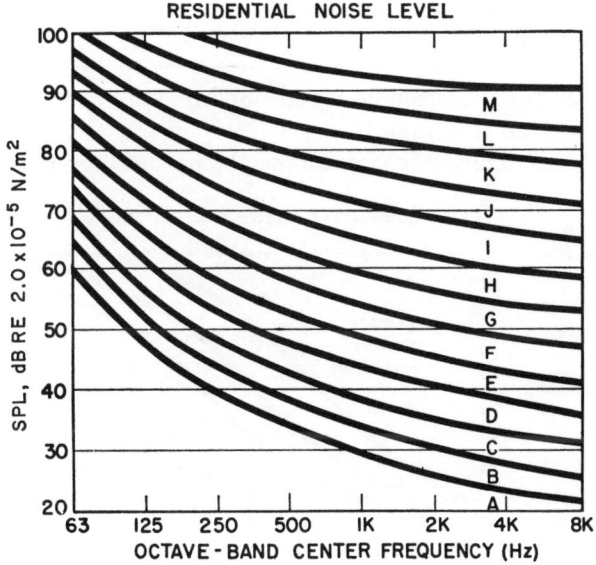

Figure 7.4

7.4) to determine the initial "level rank." The highest zone into which any of the octave band levels penetrates is the level rank of the noise.

Corrections are next applied as follows:

Condition	Correction
Pure-tone Components	+1
Wide Band Noise	0
Impulsive	+1
Nonimpulsive	0
Continuous Exposures to 1/min	0
10 to 60 Exposures/hr	−1
1 to 10 Exposures/hr	−2
4 to 20 Exposures/day	−3
1 to 4 Exposures/day	−4
1 Exposure/day	−5
Very Quiet Suburban	+1
Suburban	0
Residential Urban	−1
Urban near Some Industry	−2
Area of Heavy Industry	−3
Nighttime	0
Daytime Only	−1
No Previous Conditioning	0
Considerable Previous Conditioning	−1
Extreme Conditioning	−2

The sum of the various corrections is then applied to the original level rank to obtain the corrected level rank.

The expected community response can then be predicted by Fig. 7.5.

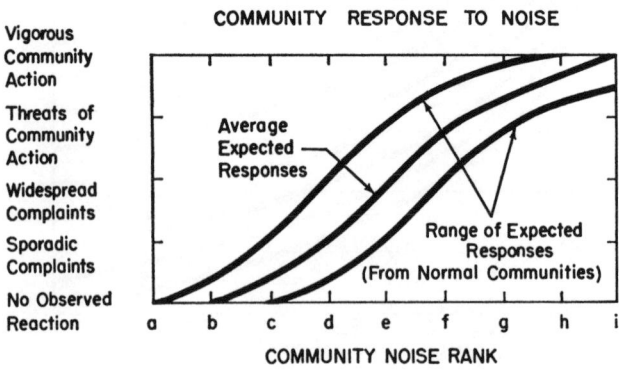

Figure 7.5

7.4 Noise Rating Numbers

In addition to speech interference level, *SIL*, a number of other rating systems have been developed to assess the annoyance and loudness of noise. Noise criteria curves (*NC*); noise rating curves (*NR*); *A*-weighted sound level (dB*A*); loudness level (*LL*); perceived noise level (*PNL*); and effective perceived noise level (*EPNL*) are some of the more common ones.

7.4.1 NC Curves. A widely used set of noise criteria for various offices, conference rooms, residences, and the like, was developed by Dr. Leo L. Beranek (Fig. 7.6). The number of the curve is numerically equal to the sound

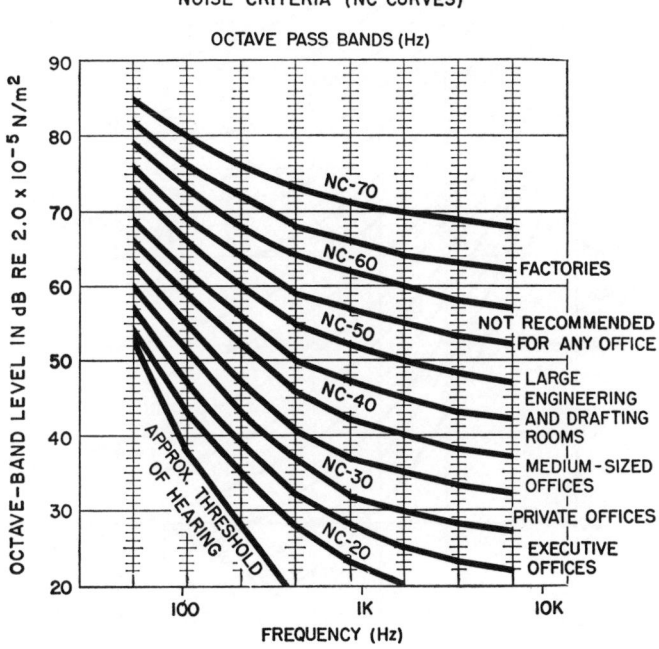

Figure 7.6

pressure level in the 1200- to 2400-Hz octave. The International Organization for Standardization, ISO, has recommended a similar set of criteria called *NR* (noise rating) curves (Fig. 7.7).

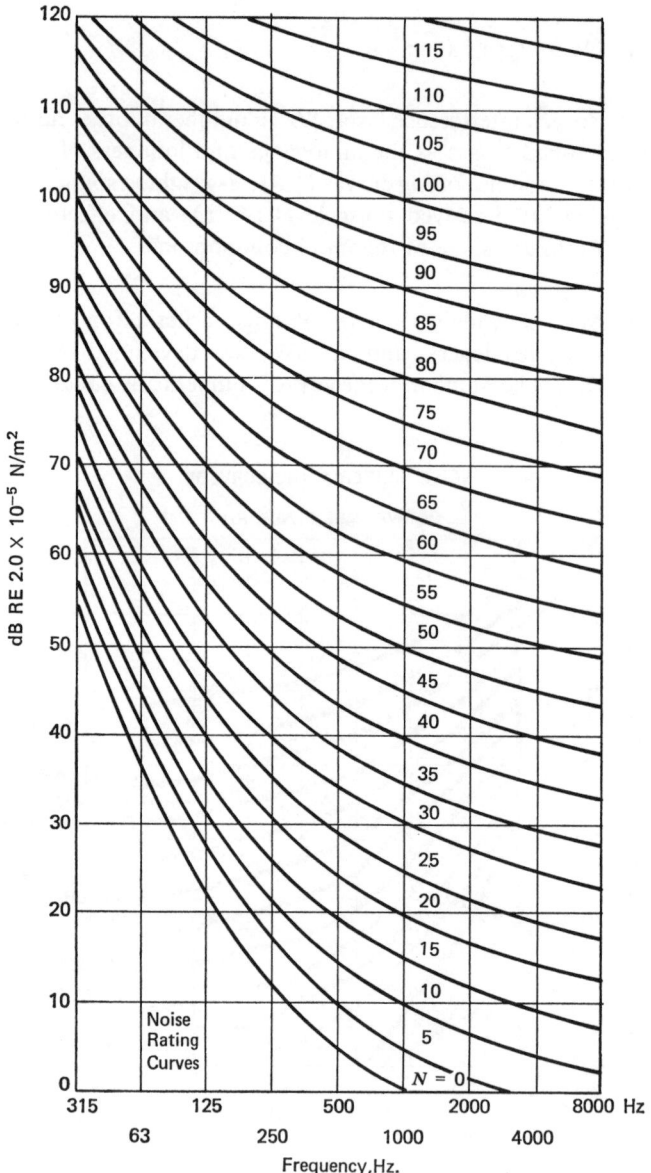

Figure 7.7

7.4.2 NEMA Criteria. The National Electrical Manufacturers Association has produced a set of noise criteria for gas turbine installations. Although this standard has been withdrawn, the curves are still used extensively by purchasers of gas turbine equipment. These curves are shown in Fig. 7.8. It

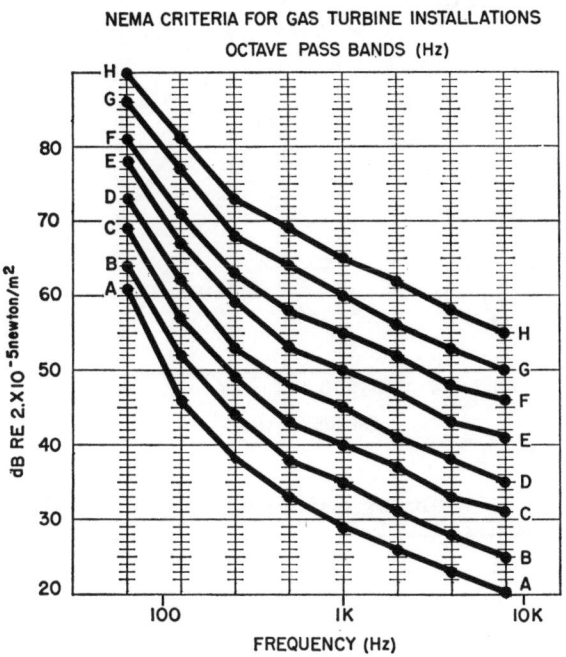

Figure 7.8

can be seen that NEMA "*A*" is approximately the same as NC25; NEMA "*B*" is about the same as NC30; NEMA "*C*" corresponds to NC35; and NEMA "*D*" is about equal to NC40. Also, if the various octave band sound pressure levels are combined, it can be seen that NEMA "*D*" would be equivalent to about 51 dB*A*; NEMA "*C*" would be equal to 49 dB*A*; NEMA "*B*" would be equal to 44 dB*A*; and NEMA "*A*" would be equal to 40 dB*A*.

7.4.3 A-Weighted Sound Level. Simplicity in using *A*-weighted sound level, or dB*A*, has resulted in its use in hearing damage criteria, speech interference levels, community annoyance criteria, regulations for vehicular traffic noise, and machinery purchase specifications.

Precautions should be taken when using dB*A* levels, especially in comparing sounds that have different intensity and frequency components. As an

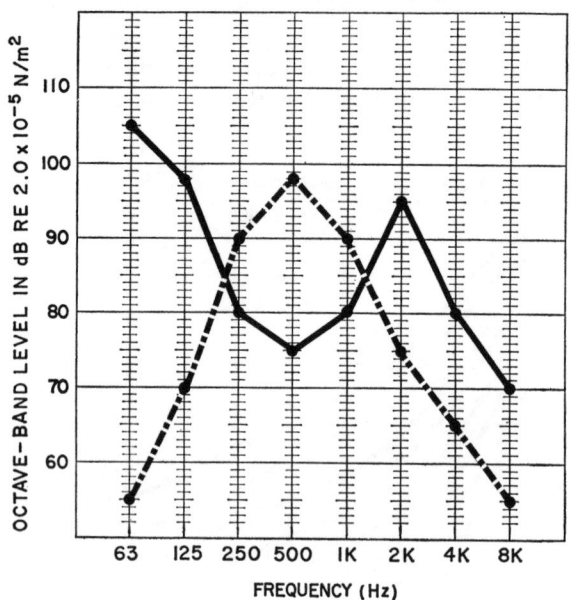

Figure 7.9

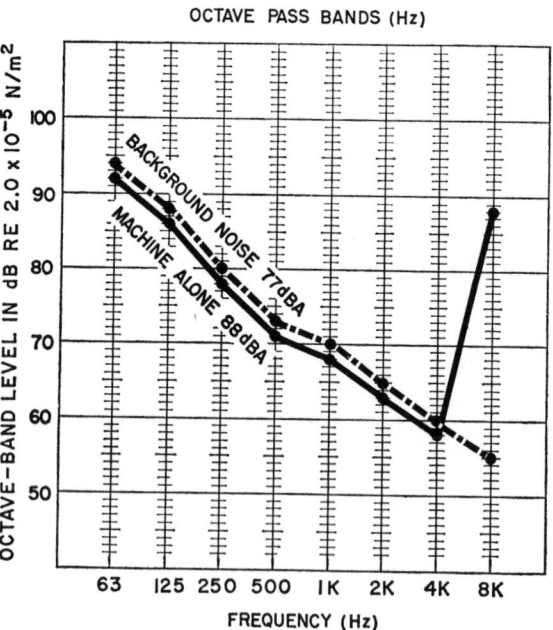

Figure 7.10

190

example, Fig. 7.9 shows octave band sound pressure levels for two noises that have the same dB*A* level as read on a sound level meter, but the distribution of the noise is entirely different in the two cases.

The dB*A* levels are inadequate for noise control work because different techniques are required to reduce low-frequency noise than those needed for high frequencies. Whenever noise control engineering is required, octave-band data are needed. In some cases, still narrower band analyses are necessary to identify the source of objectionable noise.

Other examples of erroneous conclusions that can be drawn by relying too heavily on dB*A* levels alone are shown in Figs. 7.10 and 7.11. It is usually stated that when background noise is 10 dB or more below the noise of the machine itself, the test area is satisfactory from that standpoint. Figure 7.10 shows a case where the machine noise is 88 dB*A* and the background noise is 77 dB*A*. Even though there is 11-dB*A* difference, the test data would be absolutely useless, except in the 8K-Hz octave.

Figure 7.11 shows another case where the machine alone is 88 dB*A*, and

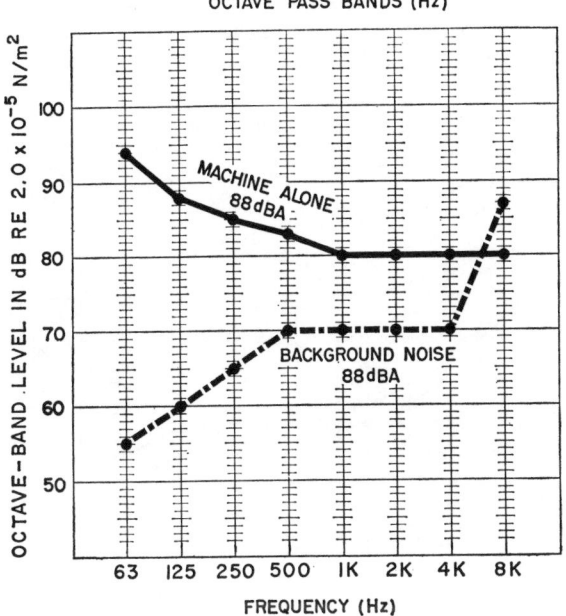

Figure 7.11

the background noise alone is also 88 dB*A*. It would ordinarily be assumed that the test data could not be used, when actually the data are perfectly valid, except in the highest octave.

Thus there is an important rule. Unless sounds have the same frequency distribution, they should be combined on an octave band basis only, and not by adding or subtracting overall levels.

7.5 Loudness Level

Loudness level, in phons, is defined as the sound pressure level of a 1000-Hz tone that sounds as loud as the noise being rated.

A set of equal-loudness contours for pure tones has been determined and published in ISO Recommendation 226, "Normal Equal-Loudness Contours for Pure Tones and Normal Threshold of Hearing under Free Field Listening Conditions." Figure 7.12 shows data taken from this recommendation. The

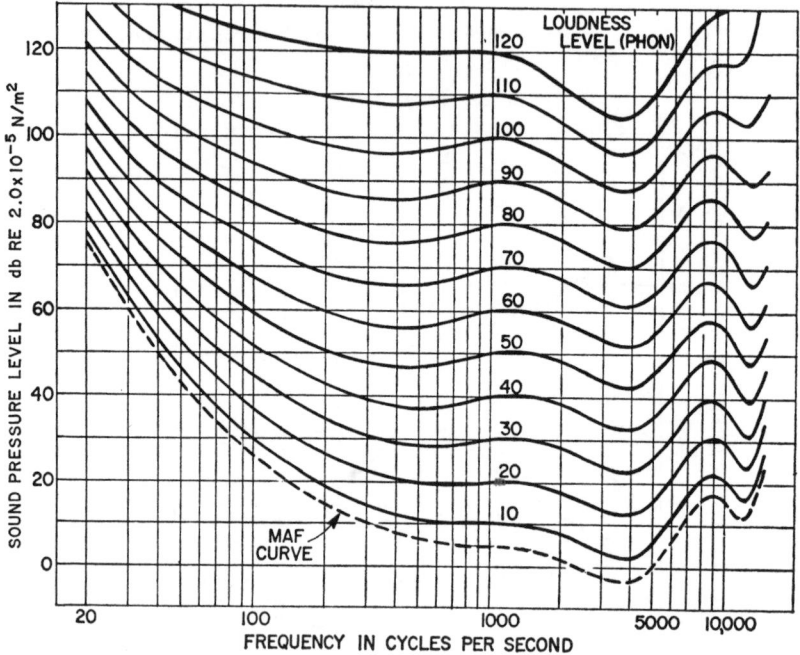

Figure 7.12

number on each curve is equal to the sound pressure level of a 1000-Hz tone that sounds as loud as all other points on that particular curve. For example, the 90-dB curve shows that 110 dB at 30 Hz, 86 dB at 500 Hz, and 80 dB at 3000 Hz all have the same loudness as a 90-dB tone at 1000 Hz. That is, they all have a loudness level of 90 phons.

Although the phon scale is logarithmic and is used to express loudness level, it does not express adequately a subjective loudness scale. That is, a sound that is twice as loud as a second sound, does not correspond to twice as many phons. Actually an increase of 10 phons corresponds to doubling the loudness.

The reason for this relationship is not fully understood. Sound pressure, in acoustics, corresponds to voltage in electrical circuits. Doubling the voltage produces a 6-dB increase. That is,

$$dB = 20 \log \frac{E_2}{E_1}$$

where E_2 and $E_1 =$ the two voltages being compared.
If $E_2 = 2E_1$

$$dB = 20 \log 2$$
$$= 20(0.301)$$
$$= 6 \text{ V}$$

With similar reasoning it would seem that doubling the loudness of a sound should represent an increase of 6 dB. But this is not the case—doubling loudness corresponds to an increase of 10 dB.

7.6 Loudness

Because it is usually not convenient to rate the loudness of sounds on the basis of phons, a loudness scale was developed in which the unit of loudness is the sone.

By definition, a 1000-Hz tone, 40 dB above a listener's threshold, produces a loudness of 1 sone. The loudness of any source that is judged by a listener to be twice as loud as that of a 1-sone tone is 2 sones. If it is n times as loud as the 1-sone tone, it has a loudness of n sones.

According to American National Standard S3.4-1968 "Procedure for the Computation of Loudness of Noise," the relation between sones and phons is given by

$$S = 2^{(P-40)/10} \qquad (7.2)$$

where $S =$ loudness, in sones, and
$\qquad P =$ loudness level in phons.
It can be seen that if the loudness level is 40 phons, $S = 1$ sone since $2^0 = 1$.

Similarly, for other levels in phons:

Phons	Sones
40	1
50	2
60	4
70	8
80	16
90	32
100	64
110	128
120	256

This shows, as was stated previously, that each time the loudness level in phons is increased 10 dB, the loudness, in sones, is multiplied by 2.

The procedure for calculating the loudness of a complex noise is based on S. S. Stevens *Procedure for Calculating Loudness: Mark VI*, and it is described in American National Standard S3.4-1968, "Procedure for the Computation of Loudness of Noise."

In the procedure it is assumed that the complex noise has been measured in terms of octave, half-octave, or third-octave bands. Each band level is first converted into a loudness index. From these the total loudness in sones is calculated by means of an empirical equation. The total loudness can then be converted into a calculated loudness level, in phons, by means of an equation, or a nomograph.

EXAMPLE 7.2   Octave band sound pressure levels measured 3 ft from an air compressor are tabulated below:
   (a) Calculate the loudness, in sones.
   (b) Calculate the loudness level, in phons.

Octave Band	SPL, dB	Loudness Index
63	87	10.0
125	103	50.0
250	112	130.0
500	102	75.0
1K	90	33.0
2K	86	30.0
4K	80	25.0
8K	74	20.0
		$\sum I = 373.0$

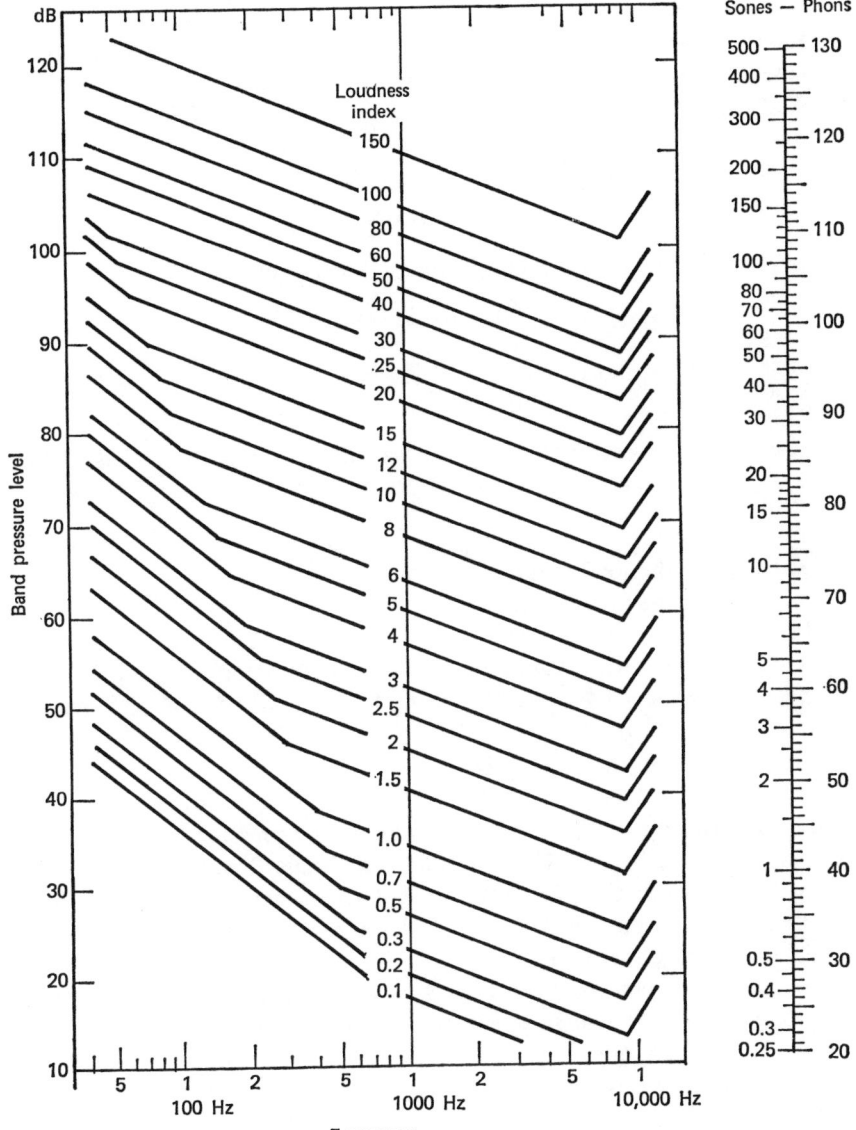

Figure 7.13 Contours of equal-loudness index.

PROCEDURE

1. First convert each octave band sound pressure level into a loudness index by means of Fig. 7.13 (taken from ANSI S3.4-1968).
2. Add the band loudness indexes to find $\sum I$. In the example this is found to be 373.0.
3. Calculate the total loudness, S_t, by the equation

$$S_t = I_m + 0.3(\sum I - I_m) \tag{7.3}$$

where I_m = the largest loudness index. Therefore,

$$S_t = 130.0 + 0.3(373.0 - 130.0)$$

$$= 130. + 73$$

$$= 203. \text{ sones}$$

4. The total loudness level, in phons, can be calculated by means of the equation

$$S_t = 2^{(P-40)/10} \tag{7.2}$$

where S_t = the total loudness, in sones, and
P = the loudness level, in phons.
From this, P is found to be 117 phons.
P can also be determined by means of the nomograph in Fig. 7.13.

7.7   Perceived Noise Level

Perceived noise level (*PNL*) is defined as the sound pressure level of a band of noise from 910 to 1090 Hz that sounds as "noisy" as the sound under comparison. It is similar to phons, except that where phons are used in equal-loudness contours, perceived noise level, expressed in *PN*dB, is used to obtain equal-annoyance contours.

The concept of perceived noise level was developed by Dr. Karl Kryter and his associates for assessing reactions to aircraft noise.

"Noisiness," in "noys," is calculated in a manner similar to that for calculating loudness, in sones. The perceived noise contributed by each frequency band is determined from a table. The total of the band contributions is then found, and it is converted to perceived noise level, in *PN*dB.

7.7.1 Effective Perceived Noise Level. Effective perceived noise level (*EPNL*) adds several additional steps to evaluate the effects of narrow band components and the duration of the noise. Narrow band noise and discrete tones have been found to be more annoying than broad band noise, and if they are present, a correction factor is added to the *PNL*. The duration of the noise is analyzed also, and suitable corrections are made depending on whether the sound being evaluated is longer or shorter than a reference duration.

Determination of effective perceived noise level requires complex instrumentation and data processing equipment.

INDEX